COORDINATE GRAPHING
FOR GRADES 6-8

BY IMMANDA BELLM

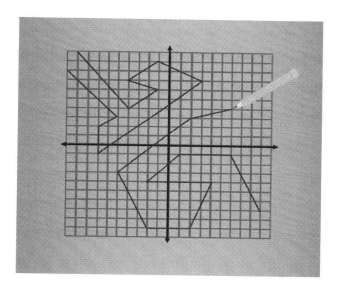

©2011 by Immanda Bellm

All rights reserved.

ISBN# 978-1466208971
ISBN-10# 146620897X

DEDICATIONS

To my husband Chris, my best friend; for helping to raise our beautiful children, for listening
to my crazy ideas, and for putting up with the late work hours and constant computer hogging.

To my two boys Ethan and Avery; you inspire me and make me laugh.

To my family and friends; for your love, support and encouragement.

Special thanks to Tracy Hartmann; my mentor and friend, who isn't afraid to tell it like it is,
is always there to catch my typos, and gives her heart and soul to improve the lives of children.

And lastly, to my students, who never fail to make each day unique and different.
Middle school is the age of humor, hope, and figuring yourself out.
Don't ever be afraid of making mistakes or trying something new. You just might surprise yourself!

TABLE OF CONTENTS

Teacher Resources	Page:
How to use this book	ii
Quadrant 1 graph paper	1
Quadrants 1&4 graph paper	2
Quadrants 1&2 graph paper	3
4-Quadrant graph paper	4
Graphing Progress Tracker	5

Section 1: Quadrant 1	6
Introduction & vocabulary page	7
Q1 Examples 1-3	8
Q1 Example Tasks 1-2	9
1. Go Green	10-11
2. Flower Power	12-13
3. Take Cover	14-15
4. Arachnophobia	16-17
5. Curved Illusion	18-19

Section 2: Quadrants 1 & 4	20
Introduction & vocabulary page	21
Q1&4 Examples 1-3	22
Q1&4 Example Tasks 1-2	23
6. Rotate-S	24-25
7. Crazy Eights	26-27
8. Inequality	28-29
9. Herbivore	30-31
10. X-Marks the Spots	32-33

Section 3: Quadrants 1 & 2	34
Introduction & vocabulary page	35
Q1&2 Examples 1-3	36
Q1&2 Example Tasks 1-2	37
11. It's a Long Shot	38-39
12. Semi-Linear	40-41
13. Man's Best Friends	42-43
14. Cube It	44-45
15. Is it "Right"?	46-47

Section 4: The Cartesian Plane	Page:
Introduction & vocabulary page	49
4-Quadrant Examples 1-3	50
4-Quadrant Example Tasks 1-2	51
16. Exploding "Pie"	52-53
17. Deep Blue	54-55
18. Midpoint Madness	56-57
19. Dream House	58-59
20. Kaleidoscope 3-D	60-61

Section 5: Transformations & Dilations	62
Introduction & vocabulary page	63
Transformation Example 1	64
Transformation Examples 2-4	65
21. Star Power	66-67
22. Star Slide	68-69
23. Star Shuffle	70-71
24. Reflecting Stars	72-73
25. Star Challenge	74-75

Section 6: Web Resources	76
Student Web Search Document	77
Cartesian Plane on the web	78
Transformations on the web	79
Optical illusions & art on the web	80
Concluding remarks	81

"Like" our Facebook page:
Coordinate Graph Art for Grades 6-8
Store Front: MathByMandy.com

Coordinate Graph Art for Grades 6-8

Growing graphing skills step by step

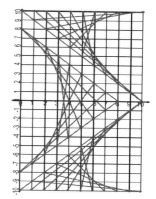

This book is was born unknowingly in my own middle school days, long before I had my own classroom or dreamed of teaching math. I would doodle during class, connecting tick-marks to make optical designs (such as the one at right). Now in my fifth year teaching middle school, I see the same desires in my own students: the need to understand, to be challenged, and to find unique ways of expressing themselves.

After various requests, "do you *any* have more graph art?" and "what else can I do for extra credit?" and most frustrating, "I could graph positives, but I can't graph negatives!!!", I had a sort of epiphany: ***There is little to no material out there to help students transition from 1 quadrant to 4 quadrants.*** So I thought, I can remedy all of these problems in one place! Thus began the journey to scaffold students from 1 quadrant, to 2 quadrants (first on the x-axis, then y-axis), and *then 4 quadrants.* And beyond.. once the basics are mastered, why not play with translations, reflections and dilations?

Read on to see some unique features that make this book both teacher- and student-friendly:

- **Student-centered language and curriculum, focused entirely on graphing.**
  No complicated instructions or assumed prior knowledge..
  Each page increases in difficulty level and states which graph paper to use.

- **Bonus instructional modules and worksheets provided for each skill section.**
  Simple, clear directions that build vocabulary and concept mastery.
  *Perfect for teacher-directed instruction/remediation **or** self-directed enrichment.*

- **Full-size picture keys immediately following the related coordinate puzzle.**
  You don't have to dig in a special section or in the back of the book.
  Student pages are immediately followed by a full-size teacher key.

- **Reproducible graph paper and progress templates in the front of the book.**
  Print a large quantity of graph paper for use with this book or on its own.
  Print copies of the progress tracker page for students to monitor their growth.

- **Unlimited copy rights within the purchaser's home or classroom.**
  Make a set for each student or keep copies on hand for quick workers.
  Purchase electronically for safe, accessible use and portability.

- Aligns with NCTM Number Sense and Geometry standards
  Provides algebra readiness in coordinate plane and spatial awareness, transformations and logical reasoning. Read more at: http://nctm.org/standards

Happy graphing!

Coordinate Graph Art for Grades 6-8

QUADRANT 1 GRAPH PAPER
FOR PICTURES 1-5 & 22

NAME _____   **DATE** _____

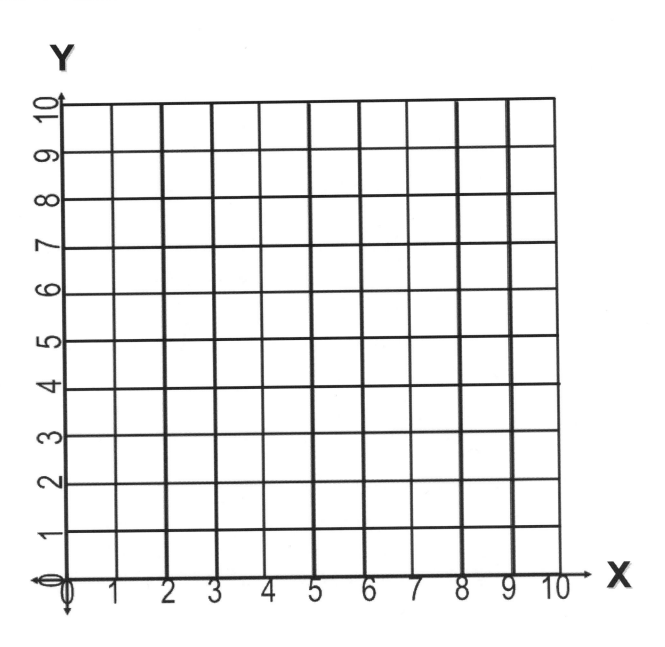

QUADRANTS 1&4 GRAPH PAPER
FOR PICTURES 6-10

NAME _____

DATE _____

QUADRANTS 1&2 GRAPH PAPER
FOR PICTURES 11-15

NAME _____ DATE _____

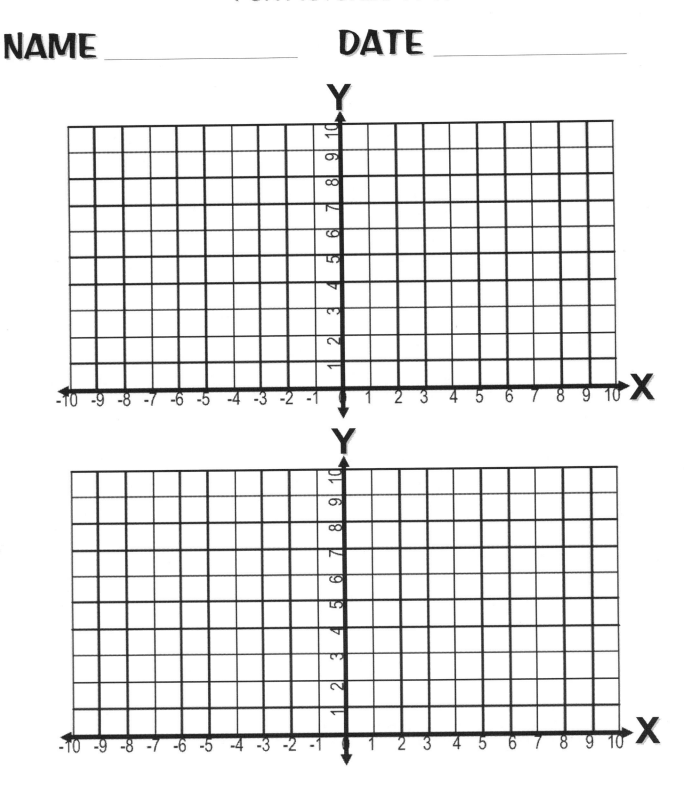

4-QUADRANT GRAPH PAPER
FOR PICTURES 16-21 & 23-25

NAME _____   DATE _____

Y

-10 -9 -8 -7 -6 -5 -4 -3 -2 -1 0 1 2 3 4 5 6 7 8 9 10 X

10 9 8 7 6 5 4 3 2 1 -1 -2 -3 -4 -5 -6 -7 -8 -9 -10

Coordinate Graph Art for Grades 6-8

GRAPHING PROGRESS TRACKER

Check off each box when you have completed the activity successfully!

NAME _____

DATE _____

Section 1: Quadrant 1   Date:

Directions & vocab pgs. 6-7 ☐ _____
Examples & practice pgs. 8-9 ☐ _____
1. Go Green ☐ _____
2. Flower Power ☐ _____
3. Take Cover ☐ _____
4. Arachnophobia ☐ _____
5. Curved Illusion ☐ _____

Section 2: Quadrants 1 & 4

Directions & vocab pgs. 20-21 ☐ _____
Examples & practice pgs. 22-23 ☐ _____
6. Rotate-S ☐ _____
7. Crazy Eights ☐ _____
8. Inequality ☐ _____
9. Herbivore ☐ _____
10. X-Marks the Spots ☐ _____

Section 3: Quadrants 1 & 2

Directions & vocab pgs. 34-35 ☐ _____
Examples & practice pgs. 36-37 ☐ _____
11. It's a Long Shot ☐ _____
12. Semi-Linear ☐ _____
13. Man's Best Friends ☐ _____
14. Cube It ☐ _____
15. Is it "Right"? ☐ _____

Section 4: The Cartesian Plane   Date:

Directions & vocab pgs. 48-49 ☐ _____
Examples & practice pgs. 50-51 ☐ _____
16. Exploding "Pie" ☐ _____
17. Deep Blue ☐ _____
18. Midpoint Madness ☐ _____
19. Dream House ☐ _____
20. Kaleidoscope 3-D ☐ _____

Section 5: Transformations & Dilations

Directions & vocab pgs. 62-63 ☐ _____
Examples & practice pgs. 64-65 ☐ _____
21. Star Power ☐ _____
22. Star Slide ☐ _____
23. Star Shuffle ☐ _____
24. Reflecting Stars ☐ _____
25. Star Challenge ☐ _____

Want more challenge?

Make your own puzzle ☐☐☐☐
Make your own transformations ☐☐☐☐
Dilate any to 1000% on sidewalk ☐☐☐☐
Shrink any to 25% or 50% ☐☐☐☐
Make a 3-D version ☐☐☐☐
Make new sidekicks for Tiny Star ☐☐☐☐
Transform Super Star into a Super
Hero with a cape and a costume! ☐☐☐☐

FINISHED EARLY? COLOR IT!

SECTION 1: THE POSITIVES
(are you positive!?)

QUADRANT 1

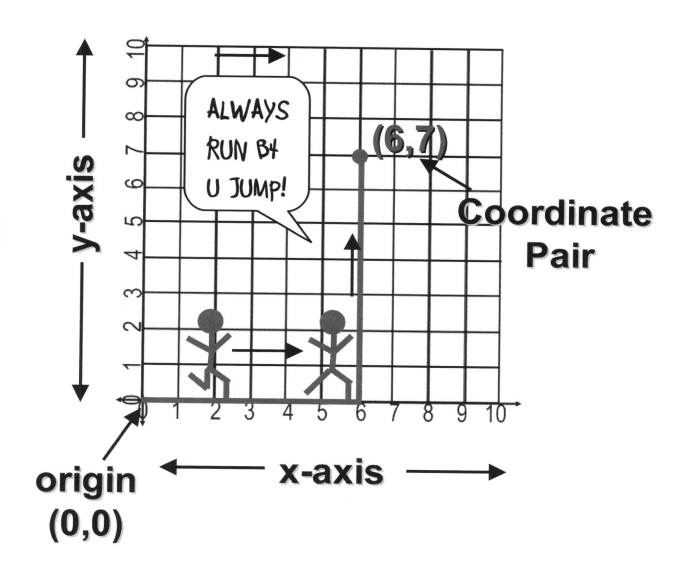

Coordinate Graph Art for Grades 6-8

QUADRANT 1

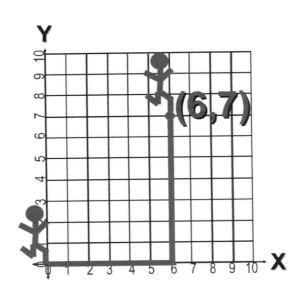

ALWAYS
RUN B4
U JUMP!

ALWAYS
RUN B4
U JUMP!

Before you "jump in" and start graphing, let's go over some important words you will see in this book (and frequently throughout the rest of your math career!). You have probably noticed some common things in the bar and line graphs you made in elementary school that will still be true now.

The horizontal part of the graph (that goes left/right) is called the **x-axis**. That's where you put the "stuff", or **variables,** that you are measuring, like time or pets.

The vertical part of the graph (that goes up/down) is called the **y-axis**. That's where you put the quantity you are measuring of your **variables**..

The center of the graph, where the **x-axis** and **y-axis** cross, is called the **origin**.. We write the **coordinate pair** for the **origin** as (0,0) because you are not going either left, right, up or down. **You <u>always</u> start from the origin when graphing.**

A **coordinate pair (X,Y)** gives directions to get you to a spot on the graph, like reading a map. The **first** half of the pair is the **x-coordinate** and the **second** half is the **y-coordinate**. It is easiest to "run" across to the x-coordinate and then "jump" to the **y-coordinate**, because you need to get a running start to jump your highest!

In this section, we are working in **Quadrant 1**. Both the **x-coordinates** and **y-coordinates** are positive. It's easy to run forwards and jump up! Right!?

Let's practice graphing coordinate pairs to familiarize you with Quadrant 1.

Example 1:

Trace with your finger how to get to **(3,9)**

The **3** is the **x-coordinate**,
which tells you to go **3 right**.

The **9** is the **y-coordinate**,
which tells you to go **9 up**.

Then name and find your own pairs with a partner.

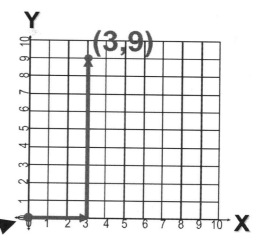

Remember, always start at the origin!

Example 2:

Can you name the **coordinate pairs** at right?

Write down your answers and check
with the correct answers below.

Hint: **Coordinates** can also land **between** the lines,
just like walking half-way down a street block.
Find the last full "block" and count forward.
(Like point **F** above... it's **between** 4 and 5!)

Example 3:

Graph the **coordinate pairs** listed below and
connect each point to the previous to make
a secret picture. If you do it right, you should
"get an A"! *Try connecting with a ruler for nice lines.*

1. (4,6)	5. (0,0)	9. (8,0)
2. (5,8)	6. (3.5,10)	10. (6.5,4)
3. (6,6)	7. (6.5,10)	11. (3.5,4)
4. (4,6)	8. (10,0)	12. (2,0)
Lift Pencil		13. (0,0)

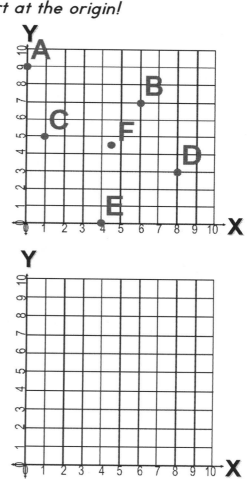

ANSWER TO EXAMPLE 3:

ANSWERS TO EXAMPLE 2:
A (0,9) B (6,7) C (1,5) D (8,3)
E (4,0) F (4.5,4.5)

9

Coordinate Graph Art for Grades 6-8

Are you ready to graph? Let's check! Find the matching coordinate pair.

Example Task 1:

This picture is MUCH less complicated to make than it looks. Each **coordinate pair** connects on a straight line to a matching pair on the other end. Your goal is to find the matching end to each pair. Use a ruler to help. *Look for cool math patterns as you work.*

For example: (0,0) connects with (10,10) Now do the rest!

(1,0) connects with (__ , __)   (8,0) connects with (__ , __)   (0,4) connects with (__ , __)
(2,0) connects with (__ , __)   (9,0) connects with (__ , __)   (0,5) connects with (__ , __)
(3,0) connects with (__ , __)   (10,0) connects with (__ , __)   (0,6) connects with (__ , __)
(4,0) connects with (__ , __)   (0,1) connects with (__ , __)   (0,7) connects with (__ , __)
(5,0) connects with (__ , __)   (0,2) connects with (__ , __)   (0,8) connects with (__ , __)
(6,0) connects with (__ , __)   (0,3) connects with (__ , __)   (0,9) connects with (__ , __)
(7,0) connects with (__ , __)

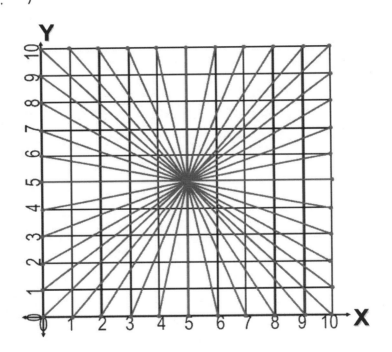

Example Task 2: Recreate the image with the **coordinate pairs** from above on your own graph paper, using a ruler. If it matches, you're ready to begin!

Example Task 1 Answers: (1,0) to (9,10); (2,0) to (8,10); (3,0) to (7,10); (4,0) to (6,10); (5,0) to (5,10); (6,0) to (4,10); (7,0) to (3,10); (8,0) to (2,10); (9,0) to (1,10); (10,0) to (0,10); (0,1) to (10,9); (0,2) to (10,8); (0,3) to (10,7); (0,4) to (10,6); (0,5) to (10,5); (0,6) to (10,4); (0,7) to (10,3); (0,8) to (10,2); (0,9) to (10,1)

10

1. GO GREEN!

Connect each coordinate pair to the previous, crossing off as you go.

(2,0)	(9,0)
(3,1)	(7,1)
(4,1)	(6,1)
(4,3)	(6,4)
(3,3)	(8,6)
(2,4)	(7,6)
(1,4)	(6,5)
(0,5)	(6,6)
(0,7)	(7,8)
(1,8)	(5,6)
(1,9)	(5,5)
(2,10)	(4,7)
(4,10)	(3,7)
(5,9)	(4,6)
(6,9)	(4,4)
(7,10)	(3,6)
(8,10)	(2,6)
(10,8)	(3,5)
(10,6)	(3,4)
(9,5)	(4,3)
(9,4)	**LIFT PENCIL**
(8,3)	**AND YOU**
(6,3)	**ARE DONE!**
LIFT PENCIL	

1. GO GREEN! KEY

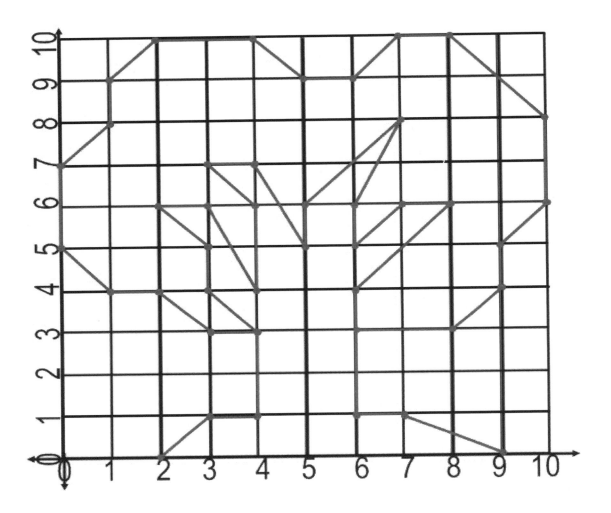

Coordinate Graph Art for Grades 6-8

2. FLOWER POWER

Connect each coordinate pair to the previous, crossing off as you go.

(1,0)	(4,2)	(8,0)	(9,5)
(0,1)	(4,5)	(10,1)	(10,6)
(1,1)	(3,5)	(10,2)	(9,6)
(1,0)	(2,6)	(9,2)	(9,7)
(2,1)	(2,8)	(8,0)	(8,6)
(2,2)	(3,7)	(8,4)	LIFT PENCIL
(1,1)	(3,6)	(7,3)	
(1,2)	(4,5)	(7,4)	(8,5)
(2,3)	(5,5)	(8,5)	(9,6)
(2,4)	(6,6)	(8,4)	LIFT PENCIL
(1,2)	(6,8)	(9,3)	AND YOU
(1,5)	(5,7)	(9,4)	ARE DONE!
(2,4)	(5,6)	(8,5)	
LIFT PENCIL	(4,5)	(7,5)	
	LIFT PENCIL	(6,4)	
(1,5)		(7,4)	
(0,4)	(3,7)	LIFT PENCIL	
(1,2)	(3,8)		
(0,3)	(4,9)	(8,5)	
(0,4)	(5,8)	(7,6)	
LIFT PENCIL	(5,7)	(6,6)	
	LIFT PENCIL	(7,5)	
(4,0)		LIFT PENCIL	
(3,2)	(8,0)		
(3,3)	(7,1)	(7,6)	
(4,2)	(6,1)	(7,7)	
(4,0)	(7,2)	(8,6)	
(5,2)	(8,1)	(8,5)	
(5,3)	LIFT PENCIL	(9,5)	
(4,2)		(10,4)	
LIFT PENCIL		(9,4)	
		LIFT PENCIL	

Use Q1 Graph Paper

2. FLOWER POWER KEY

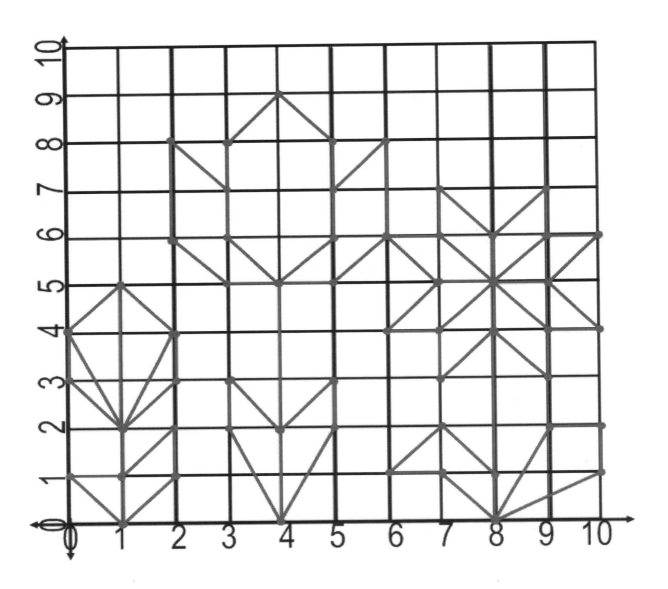

3. TAKE COVER

Connect each coordinate pair to the previous, crossing off as you go.

(7,10)
(3,7)
(5,7)
(1,4)
(3,4)
(1,0)
(6,4)
(4,4)
(8,6)
(6,6)
(10,8)
Lift Pencil

(0,6)
(0.5,6)
(1,6.5)
(2,6.5)
(2.5,7)
(2.5,7.5)
(3.5,8)
(3.5,8.5)
(4,9)
(4,9.5)
(3.5,10)
Lift Pencil

(4,0)
(4.5,0.5)
(5,0.5)
(5,1)
(5.5,1.5)
(6,1.5)
(6.5,2)
(8,2)
(8.5,1.5)
(9.5,1.5)
(10,1)
Lift Pencil

(10,2.5)
(9,2.5)
(8.5,3)
(8,3)
(7.5,3.5)
(7.5,4)
(8,4.5)
(8.5,4.5)
(8.5,5)
(9,5.5)
(9.5,5.5)
(10,5)
**Lift Pencil and
you are done!**

Use Q1 Graph Paper

15

3. TAKE COVER KEY

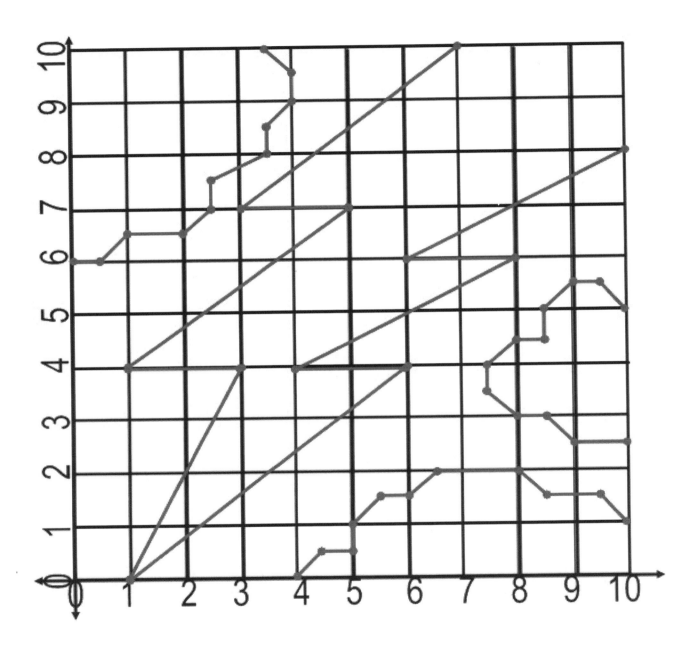

4. ARACHNOPHOBIA

Connect each coordinate pair to the previous, crossing off as you go.

You'll know what to add to the picture, if you dare!

(5,5)
(4.5,5.5)
(4.5,6)
(5,6.5)
(5.5,6.5)
(6,6)
(6,5)
(5.5,4.5)
(4.5,4.5)
(3.5,5.5)
(3.5,6.5)
(4.5,7.5)
(5.5,7.5)
(6.5,6.5)
(6.5,4.5)
(5.5,3.5)
(4,3.5)
(3,4.5)
(2.5,5.5)
(2.5,7)
(3,8)
(4,8.5)

continue to next row

(6,8.5)
(7,8)
(8,7)
(8,5)
(7.5,3.5)
(6,2.5)
(3.5,2.5)
(2,4)
(1.5,5.5)
(1.5,7.5)
(2.5,9)
(4,10)
(6,10)
(8,9)
(10,7)
(10,4)
(9,1)
(7,0)
(4,0)
(2,1)
(0,3)
(0,8)
(1,10)

Lift Pencil

Connect:
(4.5,7.5) to (3.5,10)
(5.5,7.5) to (6.5,10)
(6.5,6.5) to (10,8)
(6,5) to (10,4)
(5.5,4.5) to (7,0)
(4.5,4.5) to (2,1)
(3.5,5.5) to (0,5.5)
(3.5,6.5) to (0,8)

Lift Pencil and
you are done!

Use Q1 Graph Paper

4. ARACHNOPHOBIA KEY

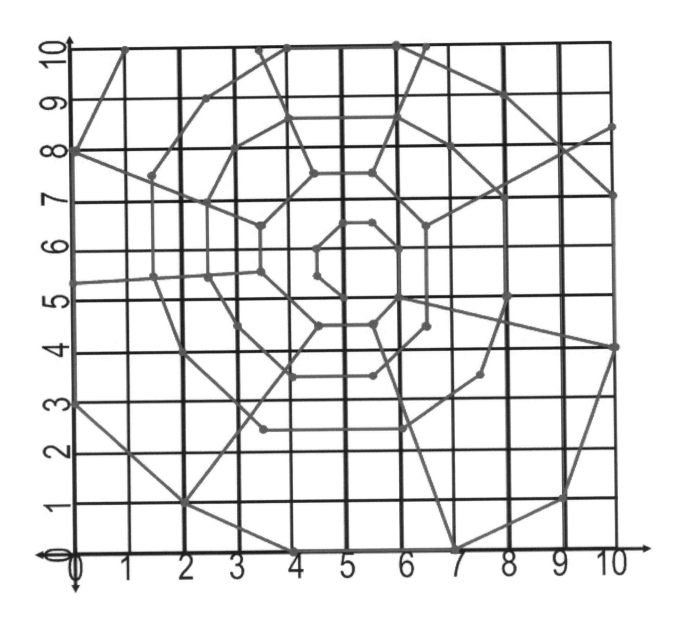

Coordinate Graph Art for Grades 6-8

5. CURVED ILLUSION

Connect each PAIR of coordinates to the other, crossing off as you go.

Use a ruler for best results.

Connect:
(1,0) to (0,10)
(2,0) to (0,9)
(3,0) to (0,8)
(4,0) to (0,7)
(5,0) to (0,6)
(6,0) to (0,5)
(7,0) to (0,4)
(8,0) to (0,3)
(9,0) to (0,2)
(10,0) to (0,1)

Connect:
(10,0) to (9,10)
(10,1) to (8,10)
(10,2) to (7,10)
(10,3) to (6,10)
(10,4) to (5,10)
(10,5) to (4,10)
(10,6) to (3,10)
(10,7) to (2,10)
(10,8) to (1,10)
(10,9) to (0,10)

Connect:
(9,1) to (1,9)
(3,3) to (7,7)

Connect:
(5.5,5.5) to (1,9)
(6,6) to (2.5,7.5)
(6.5,6.5) to (3.5,6.5)
(7,7) to (4.5, 5.5)
(4.5,4.5) to (1,9)
(4,4) to (2.5,7.5)
(3.5,3.5) to (3.5,6.5)
(3,3) to (4.5,5.5)
(4.5,4.5) to (9,1)
(4,4) to (7.5,2.5)
(3.5,3.5) to (6.5,3.5)
(3,3) to (5.5,4.5)
(5.5,5.5) to (9,1)
(6,6) to (7.5, 2.5)
(6.5,6.5) to (6.5,3.5)
(7,7) to (5.5,4.5)
Lift pencil and
you are done!

5. CURVED ILLUSION KEY

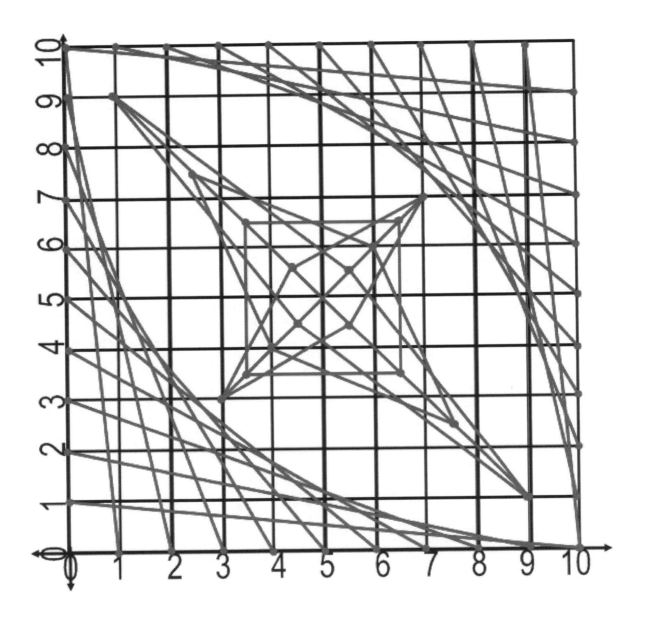

Coordinate Graph Art for Grades 6-8

SECTION 2:
POSITIVE X-COORDINATES

QUADRANTS
1&4

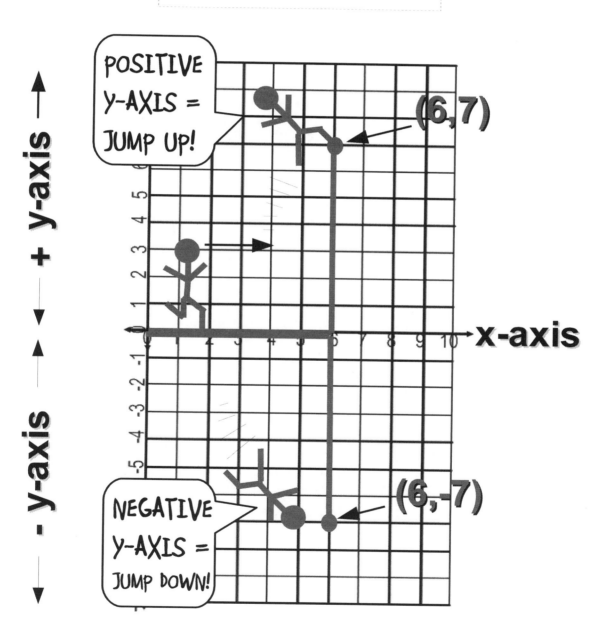

Coordinate Graph Art for Grades 6-8

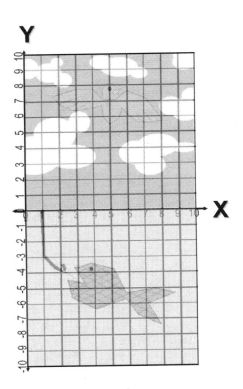

Having successfully completed Section 1, you are now an expert at graphing in positive coordinates. You have mastered the concept of graphing the **x-coordinate** first and the **y-coordinate** second. This served you well when following the pattern of running and then jumping.

In Section 2, you will now begin to start working with **negative coordinates**. We will start by introducing **negative** numbers on the **y-axis** first, that goes up and down. It makes logical sense to think about jumping "up" into the air and jumping "down" into water.

You will always need to run first, across the **x-axis**, and you will still be running to the **right**, in the **positive coordinates**. But now if you see a **negative number** for the **y-coordinate**, you will jump **down** into the water, like in a pool or lake.

Where on the picture would you go to run 3 feet, and then jump **down** 5 feet into a pool? Trace it with your finger. The **coordinate pair** for this situation is **(3, -5)**. What if you ran 6 feet and climbed **up** to a 7-foot diving board? Trace it with your finger. The **coordinate pair** for this situation would be **(6,7)** as you know. (Of course we never *really* run on pool decks, right!?)

As you work through the next section, keep in mind that the sign of the **y-coordinate** will tell you if it is "up in the air" (**Quadrant 1**) or "below the sea" (**Quadrant 4**). All **coordinate pairs** in **Quadrant 4** have the form **(X, -Y)**.

Let's practice graphing coordinate pairs in Quadrants 1 and 4.

Example 1:

Plot each **coordinate pair** on the grid at right and state if it is in **Quadrant 1 or 4**.

A. *6 right, 9 up.*   Quadrant: ___

B. *1 right, 3 down.*   Quadrant: ___

C. *10 right, 5 down.*   Quadrant: ___

Example 2:

Can you name the **coordinate pairs** at right?

Write down your answers and check with the correct answers below.

D. (__, __)   G. (__, __)   J. (__, __)
E. (__, __)   H. (__, __)   K. (__, __)
F. (__, __)   I. (__, __)   L. (__, __)

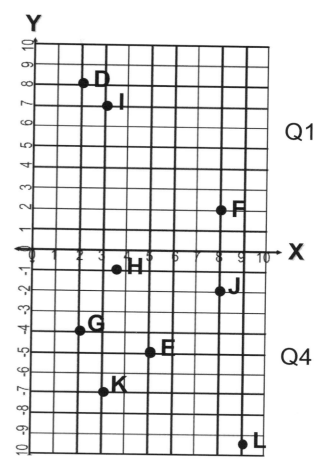

Q1

Q4

Example 3:

Graph the **coordinate pairs** in **Quadrant 4** to make a mirror image of the "A" you made in Example 3 in Section 1. What **axis** does it **reflect** over? _____

1. (4,-6)   5. (0,0)   9. (8,0)
2. (5,-8)   6. (3.5,-10)   10. (6.5,-4)
3. (6,-6)   7. (6.5,-10)   11. (3.5,-4)
4. (4,-6)   8. (10,0)   12. (2,0)
Lift Pencil   13. (0,0)

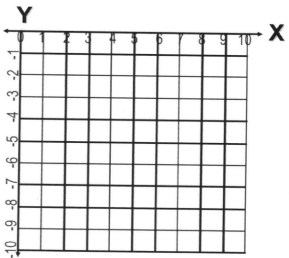

EX.3 ANSWERS: "A" should hang upside down and reflects over the x-axis.

EX.2 ANSWERS: D (2,8) E (5,-5) F (8,2) G (2,-4) H (3.5,-1) I (3,7) J (8,-2) K (3,-7) L (9,-9.5)

EX.1 ANSWERS: A (6,9) = QUADRANT 1; B (1,-3) = QUADRANT 4; C (10,-5) = QUADRANT 4

23

Before you dive in, try this mirror image out for size!

Example Task 1:

This picture is MUCH less complicated to make than it looks. Each **coordinate pair** connects on a straight line to a matching pair on the other end. Your goal is to find the matching end to each pair. Use a ruler to help. *Look for patterns of reflection as you work.*

For example: (0,0) connects with (10,2) & (10,-2)

Now you do the rest!

(10,0) connects with	(__ , __) and	(__ , __)
(0,4) connects with	(__ , __) and	(__ , __)
(0,-4) connects with	(__ , __) and	(__ , __)
(10,4) connects with	(__ , __) and	(__ , __)
(10,-4) connects with	(__ , __) and	(__ , __)
(0,8) connects with	(__ , __) and	(__ , __)
(0,-8) connects with	(__ , __) and	(__ , __)
(10,8) connects with	(__ , __) and	(__ , __)
(10,-8) connects with	(__ , __) and	(__ , __)
(4,0) connects with	(__ , __) and	(__ , __)
(6,0) connects with	(__ , __) and	(__ , __)
(4,8) connects with	(__ , __) and	(__ , __)
(4,-8) connects with	(__ , __) and	(__ , __)
(6,8) connects with	(__ , __) and	(__ , __)
(6,-8) connects with	(__ , __) and	(__ , __)

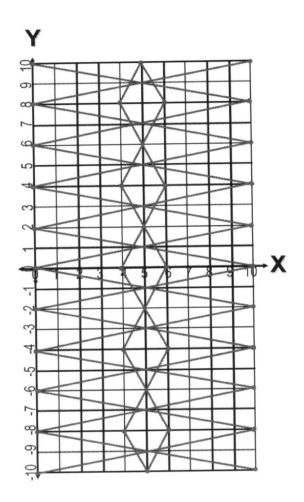

Example Task 2: Recreate the image with the **coordinate pairs** from above on your own graph paper, using a ruler. Color **reflecting** parts in the same shade so that you have a **line of symmetry** over the **x-axis**. If your design turns out similar to the one at right, you are ready to begin!

EXAMPLE TASK 1 ANSWERS: (10,0) connects with (0,2) & (0,-2); (0,4) connects with (10,2) & (10,6);
(0,-4) connects with (10,-2) & (10,-6); (10,4) connects with (0,2) & (0,6); (10,-4) connects with (0,-2) & (0,-6);
(0,8) connects with (10,6) & (10,10); (0,-8) connects with (10,-6) & (10,-10); (10,8) connects to (0,6) & 0,10);
(10,-8) connects with (0,-6) & (0,-10); (4,0) connects with (6,-4) & (6,4); (6,0) connects with (4,4) & (4,-4);
(4,8) connects with (6,4) & (6,-8); (4,-8) connects with (6,-4) & (5,-10); (6,8) connects with (4,4) & (5,10);
(6,-8) connects with (4,-4) & (5,-10).

24

6. ROTATE-S

Work with positive X-coordinates in quadrants 1 & 4.

Connect each coordinate pair in order, crossing off as you go.

Bonus points if you can name what kind of symmetry is in the picture!

(0,6)	(6,-10)
(4,10)	(2,-10)
(8,10)	(0,-6)
(10,6)	(0,-3)
(10,3)	(2,0)
(8,0)	(3,0)
(7,0)	(4,-2)
(6,2)	(3,-4)
(7,4)	(5,-7)
(5,7)	(6,-4)
(4,4)	(0,6)
(10,-6)	Lift Pencil and
Continue...	you are done!

Use Q1&4 Graph Paper

Coordinate Graph Art for Grades 6-8

6. ROTATE-S KEY

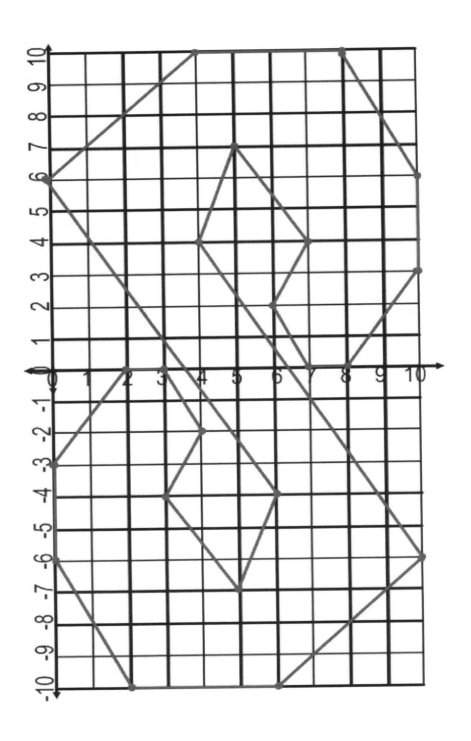

Coordinate Graph Art for Grades 6-8

7. CRAZY EIGHTS

Work with positive X-coordinates in quadrants 1 & 4.

Connect each coordinate pair in order, crossing off as you go.

(3,0)	(3,-2)	(3,6)	(4,-8)	(6,-3)
(0,3)	(2,-3)	(3,7)	(5,-7)	(6,-4)
(0,7)	(2,-7)	(4,7)	(6,-8)	(7,-4)
(3,10)	(3,-8)	(4,6)	Lift Pencil	(7,-3)
(7,10)	(7,-8)	(3,6)		(6,-3)
(10,7)	(8,-7)	Lift Pencil	(8,-6)	Lift Pencil
(10,3)	(8,-3)		(7,-5)	
(7,0)	(7,-2)	(6,6)	(8,-4)	(3,-3)
(10,-3)	(3,-2)	(6,7)	Lift Pencil	(3,-4)
(10,-7)	Lift Pencil	(7,7)		(4,-4)
(7,-10)		(7,6)	(4,-2)	(4,-3)
(3,-10)	(2,6)	(6,6)	(5,-3)	(3,-3)
(0,-7)	(3,5)	Lift Pencil	(6,-2)	
(0,-3)	(2,4)		Lift Pencil	(5,4)
(3,0)	Lift Pencil	(6,3)		(4,5)
Lift Pencil		(6,4)	(3,-6)	(5,6)
	(4,8)	(7,4)	(3,-7)	(6,5)
(3,2)	(5,7)	(7,3)	(4,-7)	(5,4)
(2,3)	(6,8)	(6,3)	(4,-6)	Lift Pencil
(2,7)	Lift Pencil	Lift Pencil	(3,-6)	
(3,8)			Lift Pencil	(5,-4)
(7,8)	(8,6)	(3,3)		(4,-5)
(8,7)	(7,5)	(3,4)	(6,-6)	(5,-6)
(8,3)	(8,4)	(4,4)	(6,-7)	(6,-5)
(7,2)	Lift Pencil	(4,3)	(7,-7)	(5,-4)
(3,2)		(3,3)	(7,-6)	Lift Pencil
Lift Pencil	(4,2)	Lift Pencil	(6,-6)	and you
	(5,3)		Lift Pencil	are done!
	(6,2)	(2,-6)		
	Lift Pencil	(3,-5)		
		(2,-4)		
		Lift Pencil		

Coordinate Graph Art for Grades 6-8

7. CRAZY EIGHTS KEY

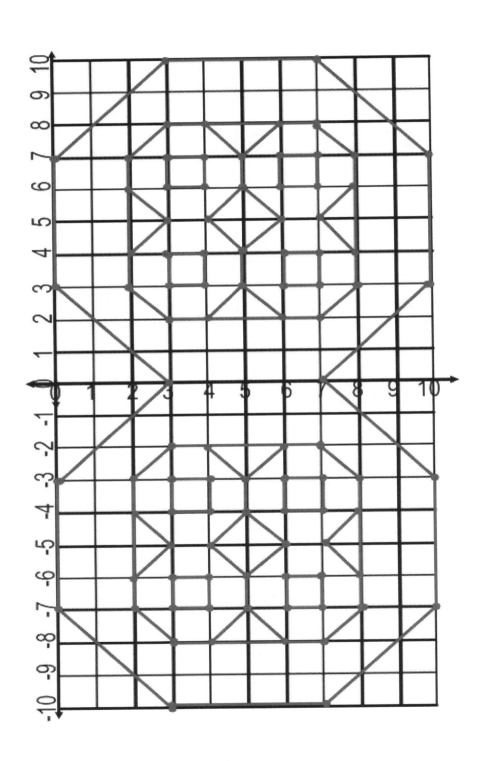

8. INEQUALITY

Work with positive X-coordinates in quadrants 1 & 4.

Connect each PAIR of coordinates to the other, crossing off as you go.

Connect:

(0,-8) to (8,0)

(8,0) to (0,8)

(0,-8) to (7,1)

(1,-7) to (6,2)

(2,-6) to (5,3)

(3,-5) to (4,4)

(4,-4) to (3,5)

(5,-3) to (2,6)

(6,-2) to (1,7)

(7,-1) to (0,8)

Connect:

(0,-10) to (10,-10)

(0,-10) to (10,0)

(1,-10) to (10,0)

(2,-10) to (9,-1)

(3,-10) to (8,-2)

(4,-10) to (7,-3)

(5,-10) to (6,-4)

(6,-10) to (5,-5)

(7,-10) to (4,-6)

(8,-10) to (3,-7)

(9,-10) to (2,-8)

(10,-10) to (1,-9)

Connect:

(0,10) to (10,10)

(0,10) to (10,0)

(1,10) to (0,10)

(2,10) to (9,1)

(3,10) to (8,2)

(4,10) to (7,3)

(5,10) to (6,4)

(6,10) to (5,5)

(7,10) to (4,6)

(8,10) to (3,7)

(9,10) to (2,8)

(10,10) to (1,9)

Use Q1&4 Graph Paper

8. INEQUALITY KEY

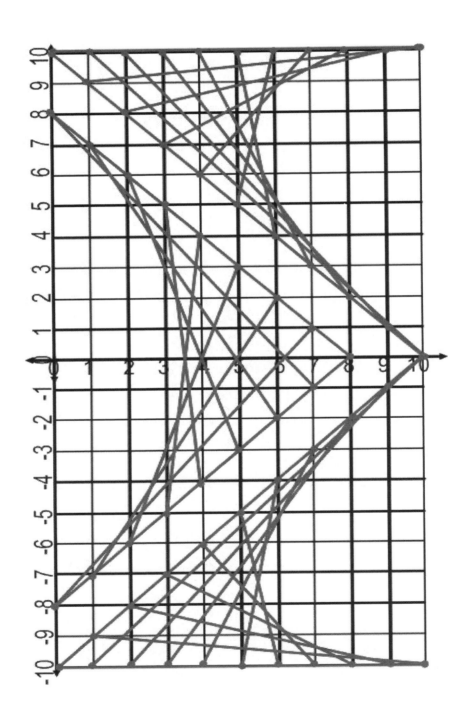

Coordinate Graph Art for Grades 6-8

9. HERBIVORE

Work with positive X-coordinates in quadrants 1 & 4.

Connect each coordinate pair in order, crossing off as you go.

(2,-4)	(7.5,6.5)	(1.5,-8)	(1,10)	(10,10)
(2,-6)	(8,7)	(1.5,-10)	(0,9)	(9,9)
(1.5,-6)	(8,7.5)	(2,-10)	(1.5,9.5)	(10,8)
(0.5,-10)	(9,7.5)	(2,-7)	(0,8)	(9,7)
(1,-10)	(9.5,8)	Lift Pencil	(0,6)	(9.5,6)
(1.5,-8)	(9,8.5)		(1,4)	(10,7)
(2,-7)	(8,8.5)	(5.5,-3)	(2,5)	Lift Pencil
(2.5,-6)	(7,8)	(5.5,-8.5)	(2.5,6.5)	
(3,-4.5)	(6,8)	(6,-10)	(2,8)	(10,5)
(3.5,-4)	(6.5,7.5)	(6.5,-10)	(2.5,9.5)	(9,5)
(3.5,-3.5)	(6,7)	(6,-8.5)	Lift Pencil	(8.5,4.5)
(4,-3.5)	(6.5,7)	(6,-4)		(8.5,3.5)
(5,-3)	(6.5,6)	Connect leg to	(2,8)	(9.5,3.5)
(6,-3)	(6,5)	body and then	(3,7)	(10,4.5)
(7,-9)	(6,4)	Lift Pencil	(3.5,8)	Lift Pencil
(7.5,-10)	(5.5,3)		(3,9.5)	and you
(8,-10)	(5,2)	(1.5,-1.5)	(4,9)	are done!
(7.5,-9)	(4.5,1)	(1,-1)	(4,10)	
(7,-6)	(4,0)	(0.5,-1)	(5,9)	
(7,-3)	(3,-1)	(1,-2)	(4,7.5)	
(7.5,-2)	(2,-1.5)	Add tail	(5,8)	
(7.5,-1)	(1.5,-1.5)	tassels	(5.5,8)	
(7,0)	(1,-2)	and then	(6,9)	
(7.5,1)	(1,-3)	Lift Pencil	(6,10)	
(7.5,2)	(1.5,-4)		(7,9)	
(7,3)	(1.5,-5)	(8.5,8)	(7,9.5)	
(7,4)	(1,-6)	(9.5,8)	(8,9)	
(7.5,5)	(1.5,-6.5)	Lift Pencil	(8.5,10)	
Continue...	Lift Pencil		Lift Pencil	

Use Q1&4 Graph Paper

9. HERBIVORE KEY

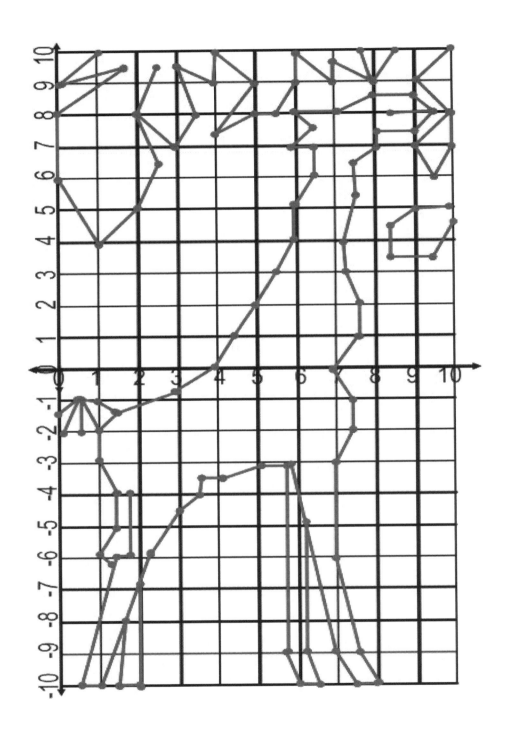

10. X MARKS THE SPOTS

Connect each PAIR of coordinates, crossing off as you go.

Connect:
(0,-10)
(10,0)
(0,10)
Lift Pencil

Connect:
(10,-10
(0,0)
(10,10)
Lift Pencil

Connect:
(0,-5)
(5,-10)
(10,-5)
(0,5)
(5,10)
(10,5)
(0,-5)
Lift Pencil

Connect:
(0,5) to (2,8)
(1,6) to (1.5,8.5)
(1.5,6.5) to (1,9)
(2,7) to (0,10)
(0,10) to (3,8)
(1,9) to (3.5,8.5)
(1.5,8.5) to (4,9)
(2,8) to (5,10)
(5,10) to (3,7)
(4,9) to (3.5,6.5)
(3.5,8.5) to (4,6)
(3,8) to (5,5)
(5,5) to (2,7)
(4,6) to (1.5,6.5)
(3.5,6.5) to (1,6)
(3,7) to (0,5)

Connect:
(0,-5) to (2,-8)
(1,-6) to (1.5,-8.5)
(1.5,-6.5) to (1,-9)
(2,-7) to (0,-10)
(0,-10) to (3,-8)
(1,-9) to (3.5,-8.5)
(1.5,-8.5) to (4,-9)
(2,-8) to (5,-10)
(5,-10) to (3,-7)
(4,-9) to (3.5,-6.5)
(3.5,-8.5) to (4,-6)
(3,-8) to (5,-5)
(5,-5) to (2,-7)
(4,-6) to (1.5,-6.5)
(3.5,-6.5) to (1,-6)
(3,-7) to (0,-5)

Connect:
(5,5) to (7,8)
(6,6) to (6.5,8.5)
(6.5,6.5) to (6,9)
(7,7) to (5,10)
(5,10) to (8,8)
(6,9) to (8.5,8.5)
(6.5,8.5) to (9,9)
(7,8) to (10,10)
(10,10) to (8,7)
(9,9) to (8.5,6.5)
(8.5,8.5) to (9,6)
(8,8) to (10,5)
(10,5) to (7,7)
(9,6) to (6.5,6.5)
(8.5,6.5) to (6,6)
(8,7) to (5,5)

Connect:
(5,-5) to (7,-8)
(6,-6) to (6.5,-8.5)
(6.5,-6.5) to (6,-9)
(7,-7) to (5,-10)
(5,-10) to (8,-8)
(6,-9) to (8.5,-8.5)
(6.5,-8.5) to (9,-9)
(7,-8) to (10,-10)
(10,-10) to (8,-7)
(9,-9) to (8.5,-6.5)
(8.5,-8.5) to (9,-6)
(8,-8) to (10,-5)
(10,-5) to (7,-7)
(9,-6) to (6.5,-6.5)
(8.5,-6.5) to (6,-6)
(8,-7) to (5,-5)

Connect:
(5,0) to (7,3)
(6,1) to (6.5,3.5)
(6.5,1.5) to (6,4)
(7,2) to (5,5)
(5,5) to (8,3)
(6,4) to (8.5,3.5)
(6.5,3.5) to (9,4)
(7,3) to (10,5)
(10,5) to (8,2)
(9,4) to (8.5,1.5)
(8.5,3.5) to (9,1)
(8,3) to (10,0)
(10,0) to (7,2)
(9,1) to (6.5,1.5)
(8.5,1.5) to (6,1)
(8,2) to (5,0)

Connect:
(5,0) to (7,-3)
(6,-1) to (6.5,-3.5)
(6.5,-1.5) to (6,-4)
(7,-2) to (5,-5)
(5,-5) to (8,-3)
(6,-4) to (8.5,-3.5)
(6.5,-3.5) to (9,-4)
(7,-3) to (10,-5)
(10,-5) to (8,-2)
(9,-4) to (8.5,-1.5)
(8.5,-3.5) to (9,-1)
(8,-3) to (10,0)
(10,0) to (7,-2)
(9,-1) to (6.5,-1.5)
(8.5,-1.5) to (6,-1)
(8,-2) to (5,0)

Connect:
(0,0) to (2,3)
(1,1) to (1.5,3.5)
(1.5,1.5) to (1,4)
(2,2) to (0,5)
(0,5) to (3,3)
(1,4) to (3.5,3.5)
(1.5,3.5) to (4,4)
(2,3) to (5,5)
(5,5) to (3,2)
(4,4) to (3.5,1.5)
(3.5,3.5) to (4,1)
(3,3) to (5,0)
(5,0) to (2,2)
(4,1) to (1.5,1.5)
(3.5,1.5) to (1,1)
(3,2) to (0,0)

Connect:
(0,0) to (2,-3)
(1,-1) to (1.5,-3.5)
(1.5,-1.5) to (1,-4)
(2,-2) to (0,-5)
(0,-5) to (3,-3)
(1,-4) to (3.5,-3.5)
(1.5,-3.5) to (4,-4)
(2,-3) to (5,-5)
(5,-5) to (3,-2)
(4,-4) to (3.5,-1.5)
(3.5,-3.5) to (4,-1)
(3,-3) to (5,0)
(5,0) to (2,-2)
(4,-1) to (1.5,-1.5)
(3.5,-1.5) to (1,-1)
(3,-2) to (0,0)

Use Q1&4 Graph Paper

10. X MARKS THE SPOTS KEY

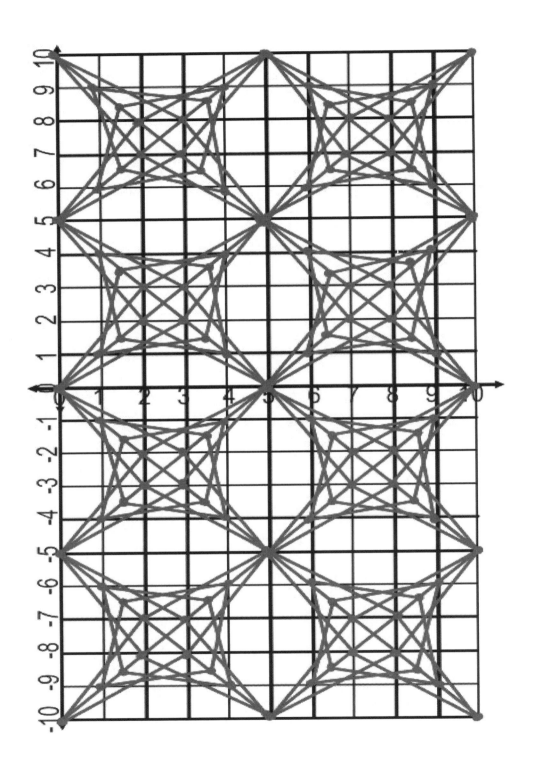

Coordinate Graph Art for Grades 6-8

SECTION 3: POSITIVE Y-COORDINATES

QUADRANTS 1&2

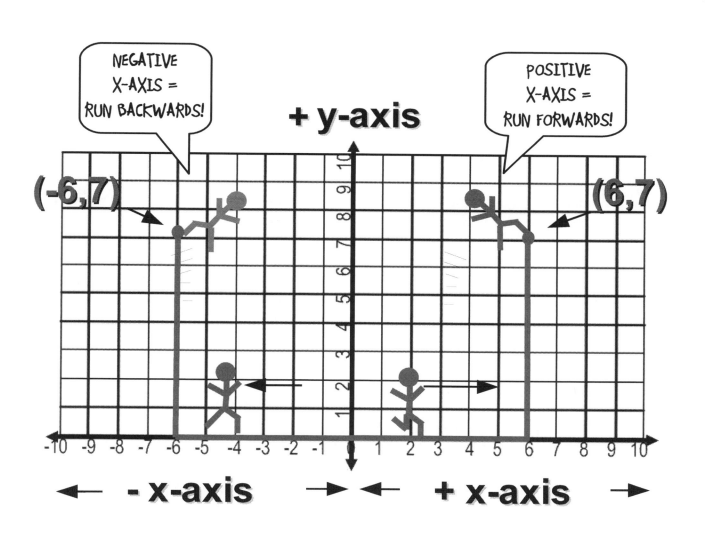

Coordinate Graph Art for Grades 6-8

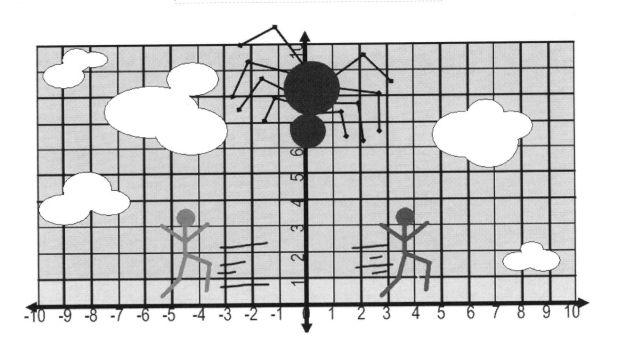

Now you have had a taste of working with **negative coordinates**. In the last section, the **negative** numbers were on the **y-axis**, that goes up and down. In this section, we will be *switching* the **negative coordinates** to the **x-axis**, that goes right and left. The **y-axis** will remain positive, so after running on the **x-axis**, you will be "jumping up".

There is a creeping spider hanging on the **y-axis** between (0,6) and (0,9.5). As you work with **negative coordinates** across the **x-axis**, think of how you would react if that spider started creeping towards you! If you are in **Quadrant 1**, on the **right**, you will run "**forwards**", to add **positive** distance between you and the spider. If you are in **Quadrant 2**, on the **left**, you will run "**backwards**", to increase the **negative** distance.

Let's say you are Mr. Green in the picture above, and the spider scares you so much that you run **backwards** 9 steps and jump **up** 5 steps. The **coordinate pair** for this situation is **(-9, 5)**. What if Mr. Green tripped instead, and only made it 2 steps **backwards** and jumped **up** one step? Trace it with your finger. The **coordinate pair** for this situation would be **(-2,1)**.

As you work through the next section, keep in mind that the sign of the **x-coordinate** will tell you if you are running **forwards** and jumping **up** into **Quadrant 1**, or running **backwards** and jumping **up** into **Quadrant 2**. The new form for **Quadrant 2** is **(-X, Y)**.

36

Let's practice graphing coordinate pairs in Quadrants 1 and 2.

Example 1:

Plot each **coordinate pair** on the grid at right and state if it is in **Quadrant 1 or 2**.

A. 3 *left*, 8 *up*.   Quadrant: ___

B. 2 *right*, 4 *up*.   Quadrant: ___

C. 10 *left*, 9 *up*.   Quadrant: ___

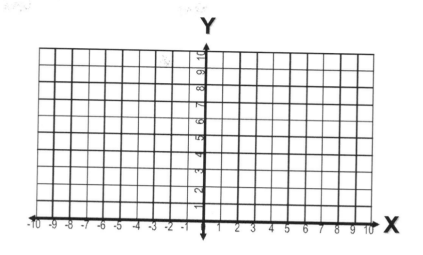

Example 2:

Can you name the **coordinate pairs** at right?

Write down your answers and check with the correct answers below.

D. (__ , __)   H. (__ , __)
E. (__ , __)   I. (__ , __)
F. (__ , __)   J. (__ , __)
G. (__ , __)   K. (__ , __)

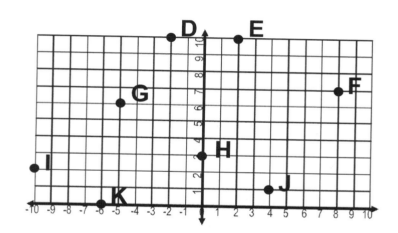

Example 3:

Graph the **coordinate pairs** in **Quadrant 2** to make a mirror image of the "A" you made in Example 3 in Section 1. What **axis** does it **reflect** over? _____

1. (-4,6)   5. (0,0)   9. (-8,0)
2. (-5,8)   6. (-3.5,10) 10. (-6.5,4)
3. (-6,6)   7. (-6.5,10) 11. (-3.5,4)
4. (-4,6)   8. (-10,0)   12. (-2,0)
Lift Pencil   13. (0,0)

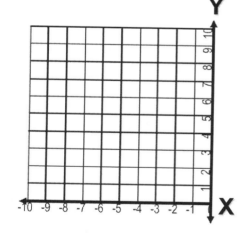

Before you dive in, try this mirror image out for size!

Example Task 1:

Write down the missing point in each **coordinate pair** in order to complete the picture shown. *Look for patterns in the **x-coordinates** as they cross over the **y-axis** from Quadrant 1 into Quadrant 2. What do you notice?*

Try this example: Connect in order: (-10,0) to (0,10) to (10,0) to make an upside-down V.

Your turn now!

Connect in order: (-10,10) to (0, __) to (__ , 10) to make a right-side up V.

Connect in order: (-10,10) to (-7.5, __) to (-10, __) to (-5, __) to (0, __) to (-2.5, __) to (0, __) to (-5, __) to (-10, __) to complete the outside of the star fruit.

Connect in order: (10,10) to (7.5, __) to (10, __) to (5, __) to (0, __) to (2.5, __) to (0, __) to (5, __) to (10, __) to complete the outside of the congruent star fruit.

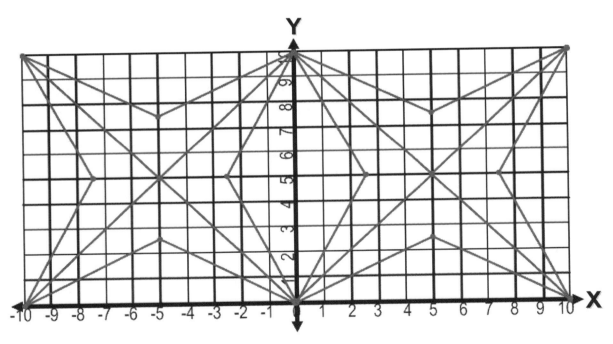

Example Task 2: Recreate the image with the **coordinate pairs** from above on your own graph paper. Color **reflecting** parts in the same shade so that you have a **line of symmetry** over the **y-axis**. If your design turns out similar to the one above, you are ready to begin!

11. IT'S A LONG SHOT

Work with positive y-coordinates in quadrants 1 & 2.

Connect each coordinate pair to the previous, crossing off as you go.

Player's Leg:
(-10,2)
(-9,2)
(-7,0)
(-6,0)
(-5,1)
(-5,2)
(-6,2)
(-7,1)
(-9,3)
(-10,3)
Lift Pencil

(-9,3)
(-9,2)
Lift Pencil

(-7,1)
(-7,0)
Lift Pencil
and add
shoe laces!

Connect:
(-4,4.5) to (-2,5.5)
(-4,3) to (-2,5)
(-2.5,3.5) to (-1.5,4.5)
(-1.5,6) to (0.5, 6)
(-1,6.5) to (-1,5.5)
(-0.5,6.5) to (-0.5,5.5)
(0,6.5) to (0,5.5)

Equipment:
(-5,0)
(-4,1)
(-3,0)
(-5,0)
Lift Pencil

(-2,6)
(-1,5)
(0,5)
(1,6)
(0,7)
(-1,7)
(-2,6)
Lift Pencil

(3,10)
(3,6)
(6,6)
(6,0)
(7,0)
(7,6)
(10,6)
(10,10)
(9,10)
(9,7)
(4,7)
(4,10)
(3,10)
Lift Pencil

Message:
(-10,6)
(-9,5)
(-8,5)
(-7,6)
(-7,7)
(-8,8)
(-9,8)
(-8,9)
(-7,9)
(-7,10)
(-8,10)
(-10,8)
(-10,6)
Lift Pencil

(-9,7)
(-9,6)
(-8,6)
(-8,7)
(-9,7)
Lift Pencil

(-6,6)
(-5,5)
(-4,5)
(-3,6)
(-3,9)
(-4,10)
(-5,10)
(-6,9)
(-6,6)
Lift Pencil

(-5,6)
(-4,6)
(-4,9)
(-5,9)
(-5,6)
Lift Pencil

(-1,4)
(0,2)
(0,0)
(1,0)
(1,2)
(2,4)
(1,4)
(0.5,3)
(0,4)
(-1,4)
Lift Pencil

(2,4)
(2,0)
(4,0)
(5,1)
(5,3)
(4,4)
(2,4)
Lift Pencil

(3,1)
(4,1)
(4,3)
(3,3)
(3,1)
Lift Pencil and
you are done!

Use Q1&2 Graph Paper

11. IT'S A LONG SHOT KEY

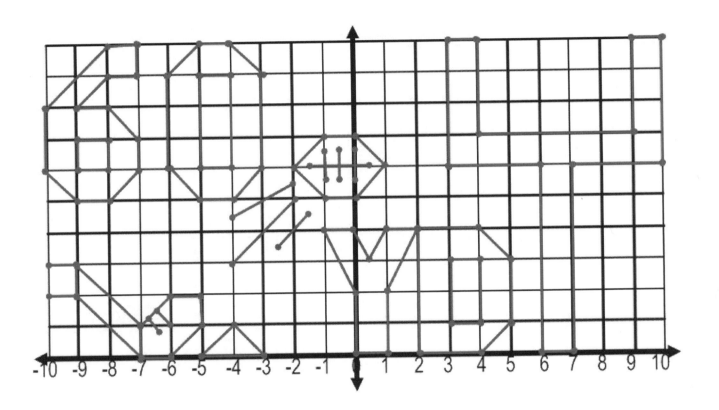

Coordinate Graph Art for Grades 6-8

12. SEMI-LINEAR

Connect each PAIR of coordinates to the other, crossing off as you go.

Use a ruler for best results.

Connect:
(-10,0) to (-9,10)
(-10,1) to (-8,10)
(-10,2) to (-7,10)
(-10,3) to (-6,10)
(-10,4) to (-5,10)
(-10,5) to (-4,10)
(-10,6) to (-3,10)
(-10,7) to (-2,10)
(-10,8) to (-1,10)
(-10,9) to (0,10)

Connect:
(10,0) to (9,10)
(10,1) to (8,10)
(10,2) to (7,10)
(10,3) to (6,10)
(10,4) to (5,10)
(10,5) to (4,10)
(10,6) to (3,10)
(10,7) to (2,10)
(10,8) to (1,10)
(10,9) to (0,10)

Connect in order:
(-7,0)
(-7,7)
(7,7)
(7,0)
Lift Pencil

Connect:
(-7,0) to (-6,7)
(-7,1) to (-5,7)
(-7,2) to (-4,7)
(-7,3) to (-3,7)
(-7,4) to (-2,7)
(-7,5) to (-1,7)
(-7,6) to (0,7)

Connect:
(7,0) to (6,7)
(7,1) to (5,7)
(7,2) to (4,7)
(7,3) to (3,7)
(7,4) to (2,7)
(7,5) to (1,7)
(7,6) to (0,7)

Connect in order:
(-4.5,0)
(-4.5,4.5)
(4.5,4.5)
(4.5,0)
Lift Pencil

Connect:
(-4.5,0) to (-4,4.5)
(-4.5,1) to (-3,4.5)
(-4.5,2) to (-2,4.5)
(-4.5,3) to (-1,4.5)
(-4.5,4) to (0,4.5)

Connect:
(4.5,0) to (4,4.5)
(4.5,1) to (3,4.5)
(4.5,2) to (2,4.5)
(4.5,3) to (1,4.5)
(4.5,4) to (0,4.5)

Connect in order:
(-3,0)
(-3,3)
(3,3)
(3,0)
Lift Pencil

Connect:
(-3,0) to (-2,3)
(-3,1) to (-1,3)
(-3,2) to (0,3)
(3,0) to (2,3)
(3,1) to (1,3)
(3,2) to 0,3)

Connect in order:
(-2,0)
(0,2)
(2,0)

Connect:
(-2,0) to (0.5,1.5)
(-1.5,0.5) to (1,1)
(-1,1) to (1.5,-0.5)
(-0.5,1.5) to (2,0)

And you are done!

12. SEMI-LINEAR KEY

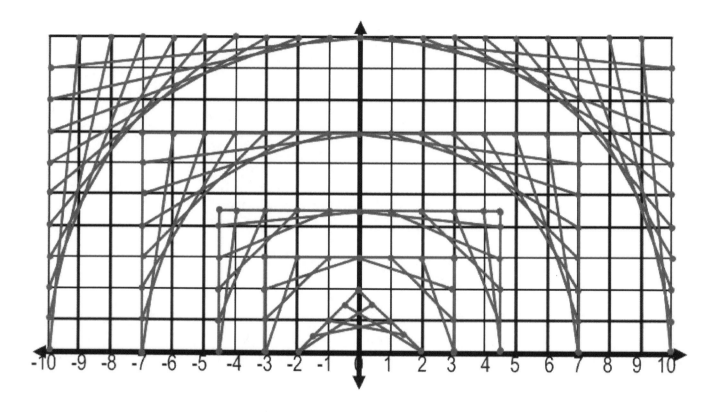

13. MAN'S BEST FRIENDS

Work with positive y-coordinates in quadrants 1 & 2.

Connect each coordinate pair to the previous, crossing off as you go.

(-10,1)	(-6,6)	(3,2)	(6.5,3)	(6.5,9)
(-9,0)	(-7,7)	(3,1)	(6.5,1.5)	(6.5,9.5)
(-5,0)	(-7,10)	(2,1.5)	(7,1)	(7,10)
(-3,1)	(-6,9)	(1.5,2.5)	(7,0.5)	(8,10)
(-2.5, 3)	(-4,9)	(1,3)	(6.5,0)	(9,9)
(-2.5, 3.5)	(-3,10)	(2,3)	Lift Pencil	(9,5)
(-1.5,4.5)	(-3,7)	(2.5,2)		(8.5,4.5)
(-3,4)	(-4,6)	(3,2)	(3.5,4.5)	(7.5,4.5)
(-3,3)	Lift Pencil	(3,3)	(2.5,4.5)	Lift Pencil
(-3.5,2)		(3.5,4.5)	(2,5)	
(-4,1.5)	(-5.5,6)	(4,5)	(2,9)	(5,6)
(-4.5,1)	(-5,6.5)	(7,5)	(3,10)	(5.5,6.5)
(-8,1)	(-5,7)	(7.5,4.5)	(4,10)	(5.5,7)
(-9,2)	(-5.5,7.5)	(8,3)	(4.5,9.5)	(5,7.5)
Lift Pencil	(-4.5,7.5)	(8,1)	(4.5,9)	(6,7.5)
	(-5,7)	(8.5,0.5)	(4,9)	(5.5,7)
(-10,1)	Lift Pencil	(8.5,0)	(3,8)	Lift Pencil
(-10,4)		(6,0)	(3,7)	
(-9,5)	(-3,1)	(5.5,1)	(4,6)	Connect:
(-7,5)	(-2.5,0.5)	(5,0)	(4,5)	(4,6.5) to (4.5,6)
(-6,6)	(-2.5,0)	(2.5,0)	Lift Pencil	(5.5,6.5) to (6,6)
(-4,6)	(-4,0)	(2.5,0.5)		(6.5,6) to (7,6.5)
(-3,5)	(-4.5,0.2)	(3,1)	(4,6)	
(-3,4)	Lift Pencil	Lift Pencil	(7,6)	Draw several dots
Lift Pencil			(8,7)	on both sides of
	Connect:	(4.5,3)	(8,8)	the dog's nose
(-7,4)	(-5,6.5) to (-4.5,6)	(4.5,1.5)	(7,9)	(above the mouth)
(-6,3)	(-6,7) to (-7.5,6.5)	(4,1)	(4.5,9)	
(-6,2)	(-6,7) to (-7.5,7)	(4,0.5)	Lift Pencil	Draw a big dot at:
(-7,1)	(-6,7) to -7.5,7.5)	(4.5,0)		(-6,8)
Lift Pencil	(-4,7) to (-2.5,6.5)	Lift Pencil	Connect:	(-4,8)
	(-4,7) to (-2.5,7)		(7,5) to (7,6)	(4,8)
	(-4,7) to (-2.5,7.5)	Connect:		(7,8)
	(-4.5,2) to (-4.5,1)	(5.5,1) to (5.5,4.5)		
				And you are done!

13. MAN'S BEST FRIENDS KEY

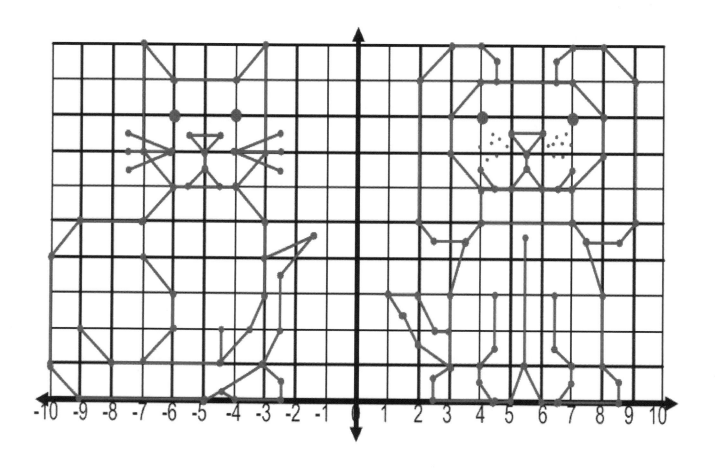

14. CUBE IT

Connect each coordinate pair to the previous, crossing off as you go.

Bonus points if you can identify the formula in the picture.

(-9,0)	(-4,2.5)	(3,4)	(-5,7)	(5,9)
(-8,0)	(-4,0.5)	(6,5.5)	(-5,6.5)	(5,6)
(-8,1)	(-5,0)	(10,5.5)	(-3,6.5)	(6,6)
(-9,1)	Lift Pencil	(7,4)	(-3,7)	(6,7)
(-9,0)		Lift Pencil	(-5,7)	(7,7)
Lift Pencil	(-3,0)		Lift Pencil	(7,6)
	(0,0)	(10,5.5)		(8,6)
(-9,1)	(0,3)	(10,1)	(-2,9)	(8,9)
(-8.5,1.5)	(-3,3)	(7,0)	(-2,6)	(7,9)
(-7.5,1.5)	(-3,0)	Lift Pencil	(0,6)	(7,8)
(-8,1)	Lift Pencil		(0,7)	(6,8)
Lift Pencil		(-9,9)	(-1,7)	(6,9)
	(-3,3)	(-8,6)	(-1,9)	(5,9)
(-7.5,1.5)	(-1,4)	(-7,6)	(-2,9)	Lift Pencil
(-7.5,0.5)	(2,4)	(-6,9)	Lift Pencil	
(-8,0)	(0,3)	(-7,9)		Draw a large
Lift Pencil	Lift Pencil	(-7.5,7.5)	(1,9)	dot like this
		(-8,9)	(1.5,6)	
(-7,0)	(2,4)	(-9,9)	(2,6)	●
(-5,0)	(2,1)	Lift Pencil	(2.5,7)	
(-5,2)	(0,0)		(3,6)	
(-7,2)	Lift Pencil	(-5,8)	(3.5,6)	between
(-7,0)		(-5,7.5)	(4,9)	the L & W
Lift Pencil	(3,0)	(-3,7.5)	(3.5,9)	and between
	(7,0)	(-3,8)	(3,7.5)	the W & H.
(-7,2)	(7,4)	(-5,8)	(2.5,8)	
(-6,2.5)	(3,4)	Lift Pencil	(2,7.5)	And you
(-4,2.5)	(3,0)		(1.5,9)	are done!
(-5,2)	Lift Pencil		(1,9)	
Lift Pencil			Lift Pencil	

Use Q1&2 Graph Paper   **45**

Coordinate Graph Art for Grades 6-8

14. CUBE IT KEY

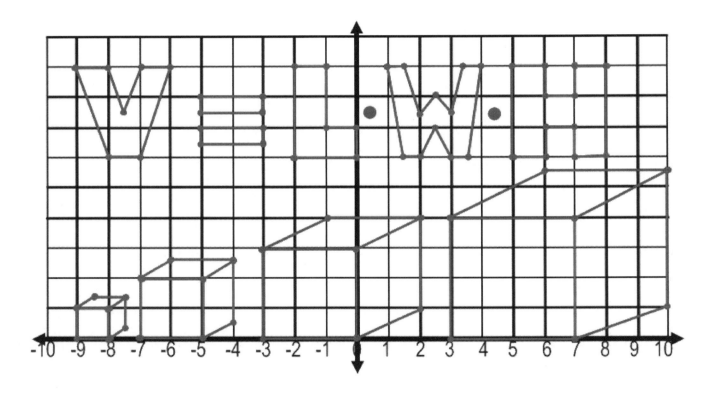

15. IS IT "RIGHT"?

Connect each coordinate pair to the previous, crossing off as you go.

Bonus points if you can name what theorem is represented!

(-4,0)	(-0.5,3)	(-8.5,6.5)	(-2,6)	(4,9)	(7.5,4)
(0,3)	(-0.5,3.5)	(-8.5,7)	(-2,10)	(2.5,8.5)	(8,3.5)
(0,0)	(0.5,3.5)	(-8,8)	(0,10)	(2.5,8)	(7,3)
(-8,0)	(0.5,3)	(-7.5,8.5)	(1,9.5)	(4,8)	(6,4)
(0,6)	Lift Pencil	(-6.5,8)	(1,8.5)	(4,7.5)	Lift Pencil
(0,3)		(-6.5,6.5)	(0,8)	(1.5,7.5)	
(-4,3)	(3.5,3)	(-5.5,7.5)	(1,7.5)	Lift Pencil	(8,1.5)
(-4,0)	(3.5,2.5)	(-5.5,7)	(1,6.5)		(8,3)
Lift Pencil	(4,2.5)	(-6.5,5.5)	(0,6)	(5,8)	(8.5,3)
	Lift Pencil	(-7,6)	(-2,6)	(5.5,8.5)	(9,2.5)
(0,0)		(-7.5,7.5)	Lift Pencil	(6.5,7.5)	(9.5,2.5)
(4,0)	(4,0.5)	(-8,7)		(6,7)	(9.5,3)
(0,3)	(4.5,0.5)	(-8,6.5)	(-1,6.5)	(5,8)	(9,3.5)
(4,3)	(4.5,0)	(-8.5,6.5)	(-1,7.5)	Lift Pencil	(9.5,3.5)
(4,0)	Lift Pencil	Lift Pencil	(0,7.5)		(10,3)
(8,0)			(0,6.5)	(6,9)	(10,2.5)
(0,6)	(-9,1)	(-4.5,7)	(-1,6.5)	(6.5,9.5)	(9.5,2)
Lift Pencil	(-10,5)	(-4.5,7.5)	Lift Pencil	(7.5,8.5)	(9,2)
	(-9,6)	(-4,7.5)		(7,8)	(8.5,2.5)
(-0.5,0)	(-5,4)	(-4,8)	(-1,8.5)	(6,9)	(8.5,1.5)
(-0.5,0.5)	(-6,3)	(-3.5,8)	(-1,9.5)	Lift Pencil	(8,1.5)
(0.5,0.5)	(-7.5,4)	(-3.5,7.5)	(0,9.5)		Lift Pencil
(0.5,0)	(-8.5,3)	(-3,7.5)	(0,8.5)	(6,4)	and you
Lift Pencil	(-8,2)	(-3,7)	(-1,8.5)	(6,5)	are done!
	(-9,1)	(-3.5,7)	Lift Pencil	(8,7)	
(-4.5,0)	Lift Pencil	(-3.5,6.5)		(9,6.5)	
(-4.5,0.5)		(-4,6.5)	(1.5,7.5)	(10,5)	
(-4,0.5)	(-9,4)	(-4,7)	(1.5,8.5)	(9,4.5)	
Lift Pencil	(-9,5)	(-4.5,7)	(3,9.5)	(8.5,5.5)	
	(-8,4.5)	Lift Pencil	(1.5,9.5)	(8,5.5)	
(-4,2.5)	(-9,4)		(2,10)	(7,4.5)	
(-3.5,2.5)	Lift Pencil		(3.5,10)	(7,4)	
(-3.5,3)			Continue...	Continue...	
Lift Pencil					

Use Q1&2 Graph Paper

Coordinate Graph Art for Grades 6-8

15. IS IT "RIGHT"? KEY

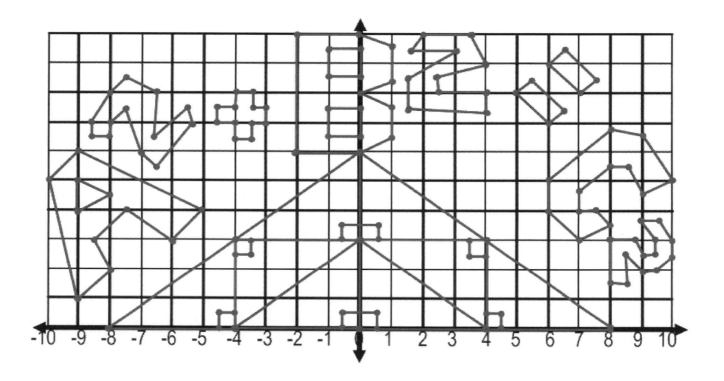

SECTION 4:
THE CARTESIAN PLANE

4 QUADRANTS

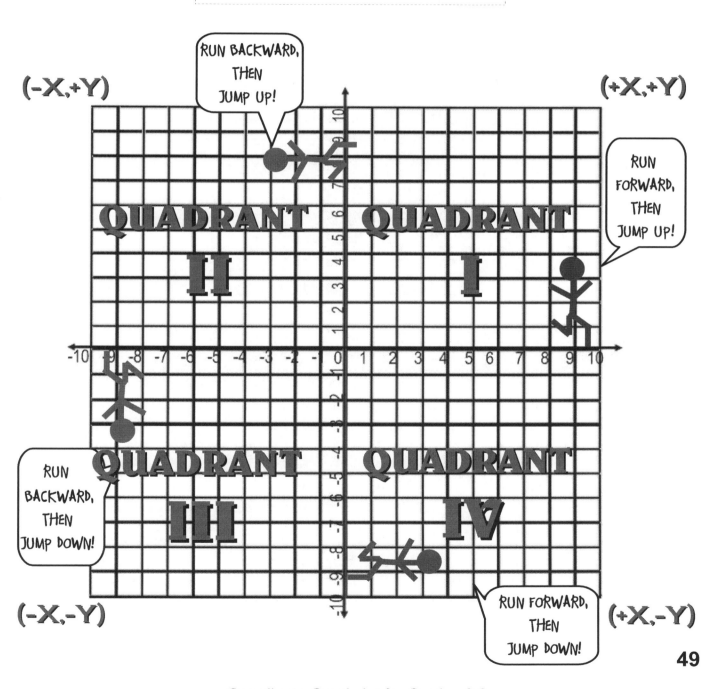

Coordinate Graph Art for Grades 6-8

THE CARTESIAN PLANE

(-X,+Y)   (+X,+Y)

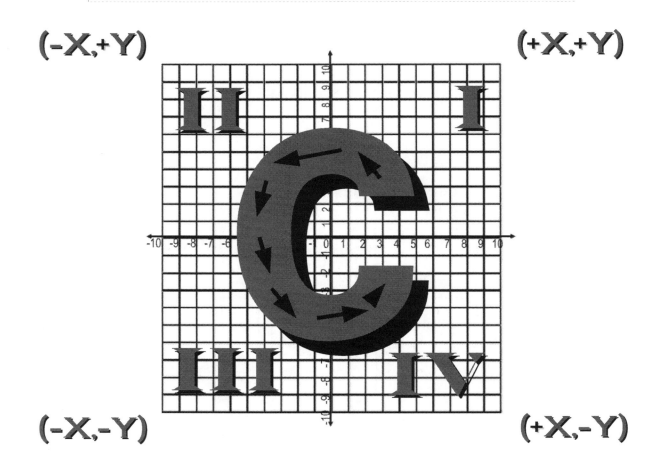

(-X,-Y)   (+X,-Y)

Congratulations on reaching the last undiscovered part of the Cartesian plane, **Quadrant 3!** Remember from previous sections that **coordinate pairs** in **Quadrant 1** have both positive numbers in the form of **(X,Y)** because you "run forwards and jump up". **Quadrant 2** has **negative** x-coordinates in the form of **(-X,Y)** because you "run backwards and then jump up". **Quadrant 4** was just the opposite, in that you "run forwards and then jump down", giving the form **(X,-Y)**. Now, in **Quadrant 3**, you will **run backwards** and **jump down**, giving the form **(-X,-Y)**. This is the **opposite** of Quadrant 1, just as Quadrants 2&4 have **opposite signs** !

Why might you ask is there a giant "C" shape on the grid? It's a memory trick, to help you remember the order and direction of the **quadrants**. While many things move clock-wise, the **quadrants** move **counter-clockwise**, in the shape and direction of writing the letter "C". Remembering that "C **is for coordinate**" will help avoid confusion. Most often you will see **Roman numerals** used to label each of the four **quadrants**..

50

Can you keep it all straight? Let's check by graphing points in all 4 quadrants.

Example 1:

Plot each **coordinate pair** on the grid at right and state which **quadrant** it appears in.

A. 9 *right*, 5 *down*.   Quadrant: ___

B. 4 *left*, 6 *up*.   Quadrant: ___

C. 1 *right*, 1 *up*.   Quadrant: ___

D. 7 *left*, 2 *down*.   Quadrant: ___

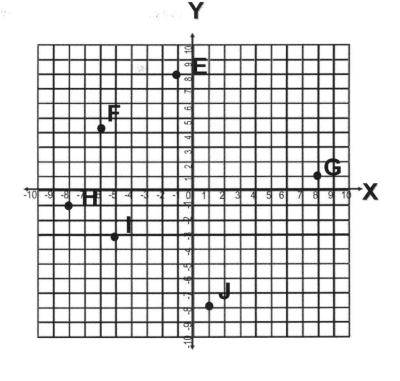

Example 2:

Can you name the **coordinate pairs**?

Write down your answers and check with the correct answers below.

E. (__, __)   H. (__, __)
F. (__, __)   I. (__, __)
G. (__, __)   J. (__, __)

Example 3:

Graph the **coordinate pairs** in **Quadrant 3** to make a **transformed** image of the "A" you made in Example 3 in Section 1. What **axes** does it **reflect** over? _____

1. (-4,-6)   5. (0,0)   9. (-8,0)
2. (-5,-8)   6. (-3.5,-10)   10. (-6.5,-4)
3. (-6,-6)   7. (-6.5,-10)   11. (-3.5,-4)
4. (-4,-6)   8. (-10,0)   12. (-2,0)
Lift Pencil   13. (0,0)

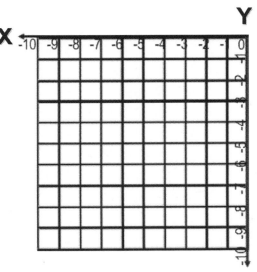

Before you dive in, try this mirror image out for size!

Example Task 1:

All four **quadrants** contain exactly the same image, except that it has been rotated **90°** in each case. In regards to the **coordinate plane**, this means that the **signs** for each **quadrant** are the only part of the **coordinate pair** to change. Use the **Quadrant 1** pairs as the basis for all others, changing the **negative** signs as applicable.

Try this example first: Graph the square starting at (5,5) by changing the signs to (-5,5), (-5,-5) & (5,-5)
                                       (Q1)   (Q2) (Q3) (Q4)

Connect in order, descending.

Then copy the pairs and **change the signs** to match the new quadrant so that the picture **rotates** around the grid as seen at right. Your turn now!

(Q1) (X,Y)	(Q2) (-X,Y)	(Q3) (-X,-Y)	(Q4) (X,-Y)
(3,2)	(-3, 2)	(-3,-2)	(3, -2)
(2,0)	(__,__)	(__,__)	(__,__)
(0,2)	(__,__)	(__,__)	(__,__)
(2,3)	(__,__)	(__,__)	(__,__)
(1,1)	(__,__)	(__,__)	(__,__)
(3,2)	(__,__)	(__,__)	(__,__)
(5,5)	(__,__)	(__,__)	(__,__)
(5,5)	(__,__)	(__,__)	(__,__)
(2,3)	(__,__)	(__,__)	(__,__)

Lift Pencil...............

(5,5)	(-5,5)	(-5,-5)	(5,-5)
(0,5)	(__,__)	(__,__)	(__,__)
(3,10)	(__,__)	(__,__)	(__,__)
(5,5)	(__,__)	(__,__)	(__,__)
(5,0)	(__,__)	(__,__)	(__,__)
(10,3)	(__,__)	(__,__)	(__,__)
(5,5)	(__,__)	(__,__)	(__,__)
(10,10)	(__,__)	(__,__)	(__,__)

Lift Pencil...............

(10,0)	(-10,0)	(-10,0)	(10,0)
(10,3)	(__,__)	(__,__)	(__,__)
(3,10)	(__,__)	(__,__)	(__,__)
(0,10)	(__,__)	(__,__)	(__,__)

Lift Pencil...............

(3,0)	(__,__)	(__,__)	(__,__)
(3,2)	(__,__)	(__,__)	(__,__)
(2,3)	(__,__)	(__,__)	(__,__)
(0,3)	(__,__)	(__,__)	(__,__)

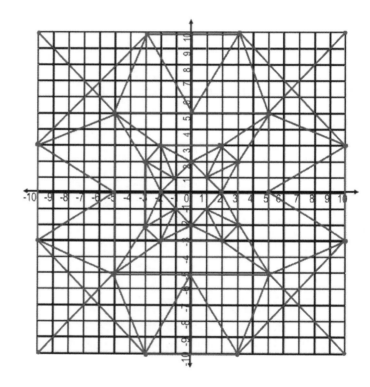

Example Task 2:

Use all 4 sets of coordinate pairs to graph your own copy of this satellite design. Color with careful attention to **patterns** that reflect over either the **x-axis, the y-axis** or **both**.

Check signs carefully and compare pictures for accuracy.

No key is provided for this task.

Coordinate Graph Art for Grades 6-8

16. EXPLODING "PIE"

Connect each coordinate pair to the previous, crossing off as you go.

(-9,0)	(-3,8)	(-5,-7)	(7,-5)	(-3,1)
(-5,3)	(-5,10)	(-7,-10)	(10,-8)	(-2,2)
(-7,8)	LIFT	LIFT PENCIL	LIFT PENCIL	(3,2)
(-2,7)	PENCIL			(3,1)
(2,10)		(1,-8)	(8,-4)	(2,1)
(3,5)	(-2,8)	(2,-10)	(10,-5)	(2,-2)
(8,6)	(-1,10)	LIFT PENCIL	LIFT PENCIL	(3,-3)
(7,1)	LIFT			(2,-3)
(10,-3)	PENCIL	(2,-8)	(9,0)	(1,-2)
(5,-4)		(5,-10)	(10,-1)	(1,1)
(6,-9)	(-7,0)	LIFT PENCIL	LIFT PENCIL	(-1,1)
(1,-7)	(-4,2)			(-1,-3)
(-3,-10)	(-5,6)	(6,-5)	(8,1)	(-2,-3)
(-4,-5)	(-2,6)	(7,-8)	(9,1)	(-2,1)
(-9,-4)	(1,8)	LIFT PENCIL	LIFT PENCIL	(-3,1)
(-6,-2)	(2,4)			LIFT PENCIL
(-9,0)	(7,5)	(-3,3)	(8,2)	AND YOU
LIFT PENCIL	(6,1)	(-4,5)	(10,5)	ARE DONE!
	(8,-2)	(-1,5)	LIFT PENCIL	
(-8,2)	(4,-3)	(0,6)		
(-10,3)	(4,-6)	(1,3)	(3,7)	
LIFT PENCIL	(1,-5)	(6,4)	(5,10)	
	(-2,-8)	(5,1)	LIFT PENCIL	
(-6,3)	(-3,-5)	(6,-1)		
(-10,6)	(-6,-3)	(4,-2)	(4,6)	
LIFT PENCIL	(-5,-2)	LIFT PENCIL	(7,9)	
	(-7,0)		LIFT PENCIL	
(-7,6)	LIFT	(3,-4)		
(-9,8)	PENCIL	(3,-5)	(5,6)	
LIFT PENCIL		(1,-4)	(10,9)	
	(-8,-2)	(-1,-5)	LIFT PENCIL	
(-4,8)	(-10,-3)	(-2,-4)		
(-7,10)	LIFT	(-3,-4)		
LIFT PENCIL	PENCIL	(-4,-3)		
		(-4,-1)		
	(-6,-6)	(-5,0)		
	(-10,-9)	LIFT PENCIL		
	LIFT			
	PENCIL			

Use 4-Quadrant Graph Paper

53

16. EXPLODING "PIE" KEY

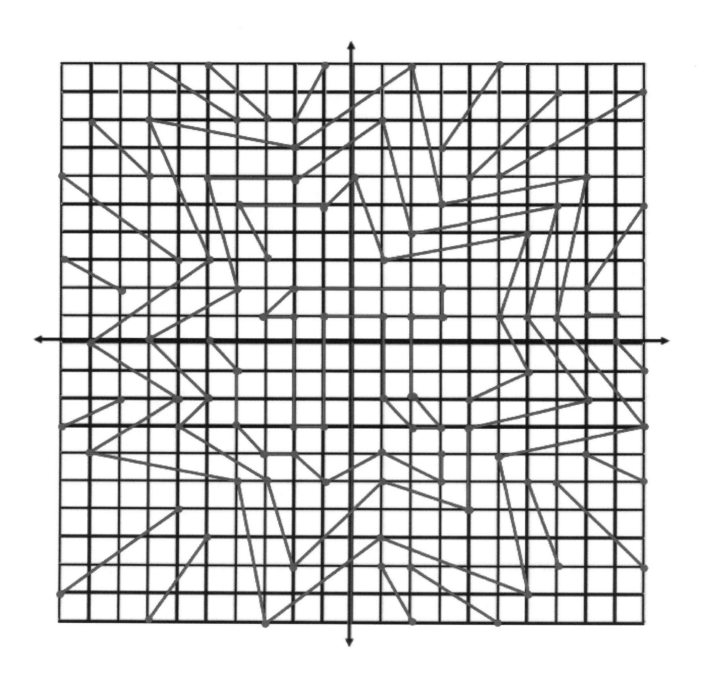

17. DEEP BLUE

Connect each coordinate pair to the previous, crossing off as you go.

Bonus points if you can name all the animal species represented!

(5,2)	(-7,2)	(-3,4)	(-1,-4)	(10,-2)
(6,3)	(-8,2)	(-1,4)	(-2,-4)	(8,-4)
(7,2)	(-8,1)	(0,1)	(-2,-5)	(9,-5)
(8,2)	(-9,3)	(-2,3)	(-3,-6)	(9,-6)
(9,3)	(-8,3)	(-3,2)	(-4,-6)	(8,-7)
(8,4)	(-7,2)	Lift Pencil	(-5,-5)	(8,-8)
(7,4)	(-6,1)		(-7,-4)	(7,-9)
(6,3)	(-4,1)	(-7,-4)	Lift Pencil	(8,-10)
(5,4)	(-3,2)	(-6,-5)		(8,-9)
(5,2)	(-3,4)	(-4,-8)	(-10,-2)	(9,-8)
Lift Pencil	(-5,6)	(-5,-10)	(-9,-4)	(9,-7)
	(-7,6)	(-4,-9)	(-9,-5)	(10,-6)
(-1,7)	(-8,5)	(-3,-7)	(-10,-6)	(10,-5)
(0,6)	(-8,3)	(-2,-7)	(-10,-7)	(9,-4)
(2,5)	Lift Pencil	(-2,-8)	(-9,-8)	(10,-2)
(4,5)		(-1,-7)	(-9,-9)	Lift Pencil
(5,6)	(-5,6)	(0,-7)	(-8,-10)	and you
(6,5)	(-5,7)	(3,-5)	(-8,-8)	are done!
(6,8)	(-3,6)	(3,-8)	(-9,-7)	
(5,7)	(-4,6)	(4,-7)	(-9,-6)	
(4,9)	(-4,5)	(5,-5)	(-8,-5)	
(2,9)	Lift Pencil	(5,-4)	(-8,-4)	
(0,8)		(6,-3)	(-10,-2)	
(-1,7)	(-6,1)	(7,-2)	Lift Pencil	
Lift Pencil	(-6,-1)	(8,0)		
	(-4,-2)	(6,0)	(5,-7)	
(-8,5)	(-5,0)	(4,-1)	(5,-10)	
(-9,6)	(-4,1)	(3,-1)	(6,-9)	
(-9,7)	Lift Pencil	(1,0)	(6,-8)	
(-8,7)		(1,-3)	(5,-7)	
(-7,6)		Continue...	Lift Pencil	
Lift Pencil				

17. DEEP BLUE KEY

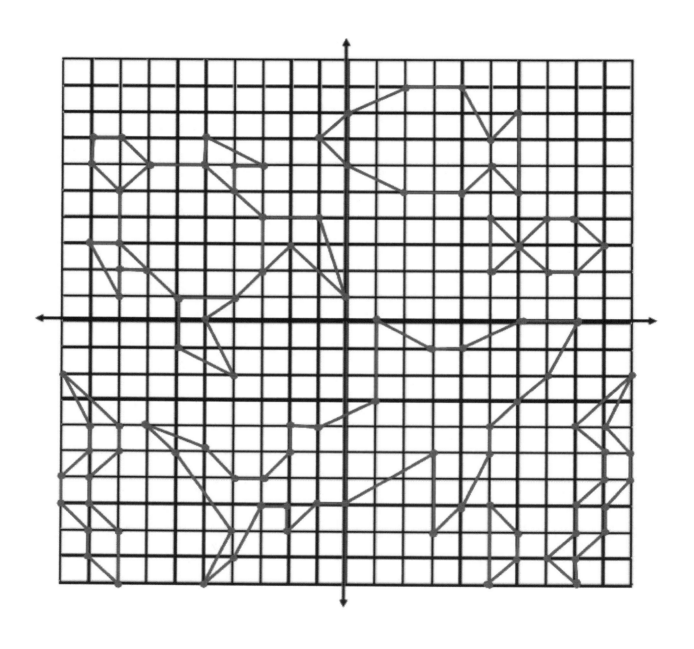

Coordinate Graph Art for Grades 6-8

18. MIDPOINT MADNESS

Connect each coordinate pair to the previous, crossing off as you go.

Color each layer of embedded triangles in a different color for a neat effect.

(-10,10)	(-5,0)	(-1,1)	(-0.5,0)
(10,10)	(0,5)	(1,1)	(0,0.5)
(10,-10)	(5,0)	(1,-1)	(0.5,0)
(-10,-10)	(0,-5)	(-1,-1)	(0,-0.5)
(-10,10)	(-5,0)	(-1,1)	(-0.5,0)
Lift Pencil	Lift Pencil	Lift Pencil	Lift Pencil
			and you
(-10,0)	(-2.5,2.5)	(-1,0)	are done!
(0,10)	(2.5,2.5)	(0,1)	
(10,0)	(2.5,-2.5)	(1,0)	
(0,-10)	(-2.5,-2.5)	(0,-1)	
(-10,0)	(-2.5,2.5)	(-1,0)	
Lift Pencil	Lift Pencil	Lift Pencil	
(-5,5)	(-2.5,0)	(-0.5,0.5)	
(5,5)	(0,2.5)	(0.5,0.5)	
(5,-5)	(2.5,0)	(0.5,-0.5)	
(-5,-5)	(0,-2.5)	(-0.5,-0.5)	
(-5,5)	(-2.5,0)	(-0.5,0.5)	
Lift Pencil	Lift Pencil	Lift Pencil	

Use 4-Quadrant Graph Paper

57

18. MIDPOINT MADNESS KEY

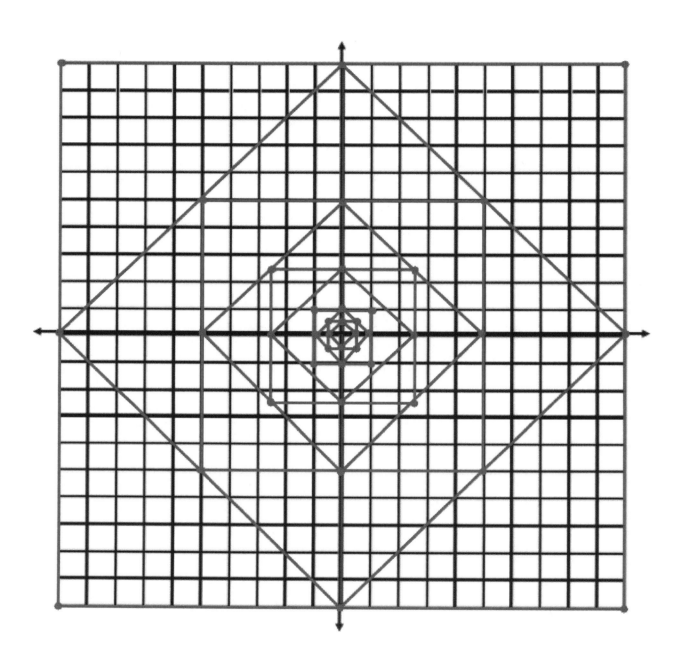

19. DREAM HOUSE

Connect each coordinate pair to the previous, crossing off as you go.

(-10,-7)	(-4,-2)	(8,2)	(0,-2)	(-7.5,10)
(10,-7)	(-3,-2)	(8,-7)	(-1,-1)	(-7.5,9.5)
Lift Pencil	(-3,-3)	Lift Pencil	(-1,0)	(-6.5,8.5)
	(-4,-3)		(0,1)	(-5.5,8.5)
(-8,-7)	(-4,-2)	(2,-7)	(1,1)	(-5,8)
(-8,-1)	Lift Pencil	(2,-4)	(2,0)	(-4,8)
(-9,-1)		(5,-4)	(2,-1)	(-3.5,7.5)
(-5,5)	(-8,0.5)	(5,-7)	(1,-2)	(-2.5,7.5)
(-1,-1)	(-8,4)	Lift Pencil	(0,-2)	(-1.5,7)
(-8,-1)	(-7,4)		(0,1)	(-0.5,7)
Lift Pencil	(-7,2)	(6,6)	Lift Pencil	(0,7.5)
	Lift Pencil	(6,7)		(0.5,7.5)
(-6,-7)		(8,7)	Connect:	(1,8)
(-6,-4)	(-10,6)	(8,4)	(1,1) to (1,-2)	(1,8.5)
(-4,-4)	(-9,5.5)	Lift Pencil	(-1,0) to (2,0)	(2,8.5)
(-4,-7)	(-8.5,5)		(-1,-1) to (2,-1)	(2.5,9)
Lift Pencil	(-8,5)	(-1,-4)		(2.5,9.5)
	(-7.5,4)	(1,-4)	(4,-0.5)	(2,10)
(-2,-1)	(-7,5)	(1,-6)	(4.5,0.5)	Lift Pencil
(-2,-7)	(-7,6)	(-1,-6)	(5,1)	
Lift Pencil	(-7.5,6.5)	(-1,-4)	(6,1)	(8,10)
	(-8.5,6.5)	Lift Pencil	(6.5,0.5)	(7,9)
(-7,-2)	(-9,7)		(7,-0.5)	(6.5,9)
(-6,-2)	(-9,8)	Connect:	(7,-3)	(6,8)
(-6,-3)	(-9.5,8.5)	(0,-4) to (0,-6)	(4,-3)	(7,7)
(-7,-3)	(-10,8.5)	(-1,-5) to (1,-5)	(4,-0.5)	(7,8)
(-7,-2)	Lift Pencil			(7.5,8.5)
Lift Pencil			Connect:	(8.5,8.5)
	(-3,2)		(5,1) to (5,-3)	(9,9)
	(3.5,9)		(6,1) to (6,-3)	(9.5,10)
	(10,2)		(4,-0.5) to (7,-0.5)	Lift Pencil
	(-3,2)		(4,-2) to (7,-2)	and you are done!
	Lift Pencil			

19. DREAM HOUSE KEY

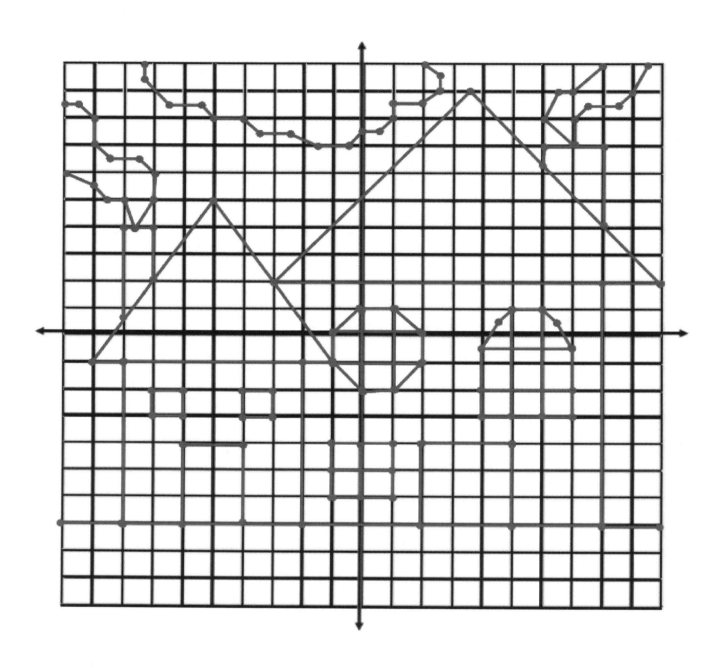

20. KALEIDOSCOPE 3-D

Connect each PAIR of coordinates, crossing off as you go. Use a ruler.

Bonus points if you can name the scale factor between the similar figures!

Connect:
(-10,0) to (0,1)
(-9,0) to (0,2)
(-8,0) to (0,3)
(-7,0) to (0,4)
(-6,0) to (0,5)
(-5,0) to (0,6)
(-4,0) to (0,7)
(-3,0) to (0,8)
(-2,0) to (0,9)
(-1,0) to (0,10)

Connect:
(10,0) to (0,1)
(9,0) to (0,2)
(8,0) to (0,3)
(7,0) to (0,4)
(6,0) to (0,5)
(5,0) to (0,6)
(4,0) to (0,7)
(3,0) to (0,8)
(2,0) to (0,9)
(1,0) to (0,10)

Connect:
(-10,0) to (0,-1)
(-9,0) to (0,-2)
(-8,0) to (0,-3)
(-7,0) to (0,-4)
(-6,0) to (0,-5)
(-5,0) to (0,-6)
(-4,0) to (0,-7)
(-3,0) to (0,-8)
(-2,0) to (0,-9)
(-1,0) to (0,-10)

Connect:
(10,0) to (0,-1)
(9,0) to (0,-2)
(8,0) to (0,-3)
(7,0) to (0,-4)
(6,0) to (0,-5)
(5,0) to (0,-6)
(4,0) to (0,-7)
(3,0) to (0,-8)
(2,0) to (0,-9)
(1,0) to (0,-10)

Connect:
(-6,2) to (-6,10)
(-10,6) to (-2,6)
(6,2) to (6,10)
(10,6) to (2,6)
(-6,-2) to (-6,-10)
(-10,-6) to (-2,-6)
(6,-2) to (6,-10)
(10,-6) to (2,-6)

Connect:
(-6,2) to (-7,6)
(-6,3) to (-8,6)
(-6,4) to (-9,6)
(-6,5) to (-10,6)
(-6,7) to (-10,6)
(-6,8) to (-9,6)
(-6,9) to (-8,6)
(-6,10) to (-7,6)

Connect:
(-6,2) to (-5,6)
(-6,3) to (-4,6)
(-6,4) to (-3,6)
(-6,5) to (-2,6)
(-6,7) to (-2,6)
(-6,8) to (-3,6)
(-6,9) to (-4,6)
(-6,10) to (-5,6)

Connect:
(6,2) to (7,6)
(6,3) to (8,6)
(6,4) to (9,6)
(6,5) to (10,6)
(6,7) to (10,6)
(6,8) to (9,6)
(6,9) to (8,6)
(6,10) to (7,6)

Connect:
(6,2) to (5,6)
(6,3) to (4,6)
(6,4) to (3,6)
(6,5) to (2,6)
(6,7) to (2,6)
(6,8) to (3,6)
(6,9) to (4,6)
(6,10) to (5,6)

Connect:
(-6,-2) to (-7,-6)
(-6,-3) to (-8,-6)
(-6,-4) to (-9,-6)
(-6,-5) to (-10,-6)
(-6,-7) to (-10,-6)
(-6,-8) to (-9,-6)
(-6,-9) to (-8,-6)
(-6,-10) to (-7,-6)

Connect:
(-6,-2) to (-5,-6)
(-6,-3) to (-4,-6)
(-6,-4) to (-3,-6)
(-6,-5) to (-2,-6)
(-6,-7) to (-2,-6)
(-6,-8) to (-3,-6)
(-6,-9) to (-4,-6)
(-6,-10) to (-5,-6)

Connect:
(6,-2) to (7,-6)
(6,-3) to (8,-6)
(6,-4) to (9,-6)
(6,-5) to (10,-6)
(6,-7) to (10,-6)
(6,-8) to (9,-6)
(6,-9) to (8,-6)
(6,-10) to (7,-6)

Connect:
(6,-2) to (5,-6)
(6,-3) to (4,-6)
(6,-4) to (3,-6)
(6,-5) to (2,-6)
(6,-7) to (2,-6)
(6,-8) to (3,-6)
(6,-9) to (4,-6)
(6,-10) to (5,-6)

Use 4-Quadrant Graph Paper

Coordinate Graph Art for Grades 6-8

20. KALEIDOSCOPE 3-D KEY

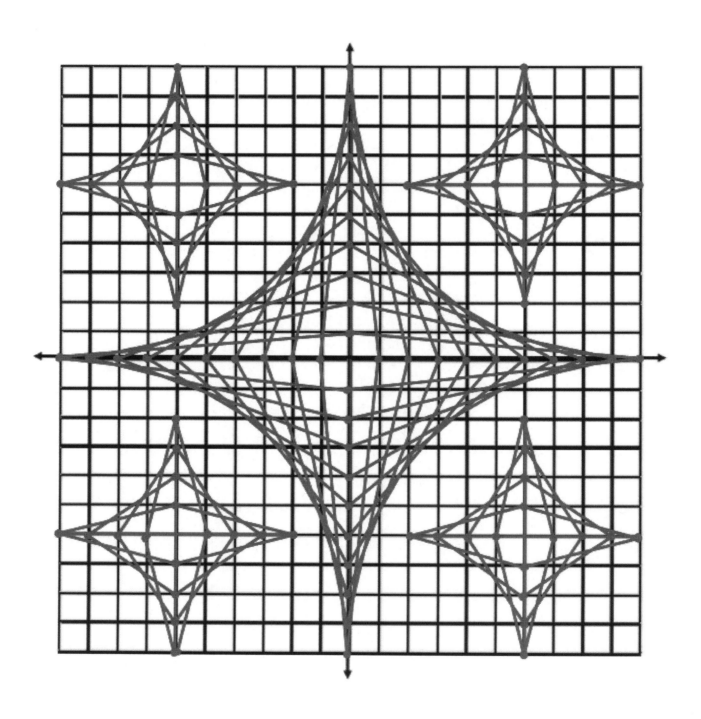

Coordinate Graph Art for Grades 6-8

SECTION 5:
TRANSFORMATIONS & DILATIONS

IN THE COORDINATE PLANE

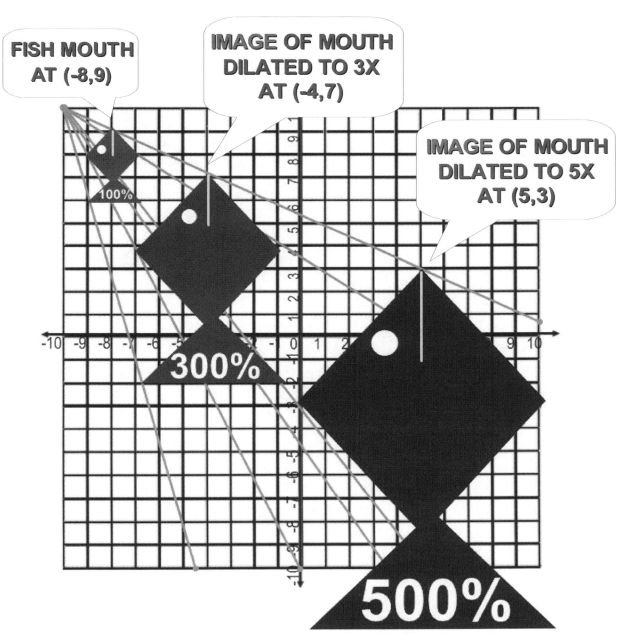

FISH MOUTH AT (-8,9)

IMAGE OF MOUTH DILATED TO 3X AT (-4,7)

IMAGE OF MOUTH DILATED TO 5X AT (5,3)

100%

300%

500%

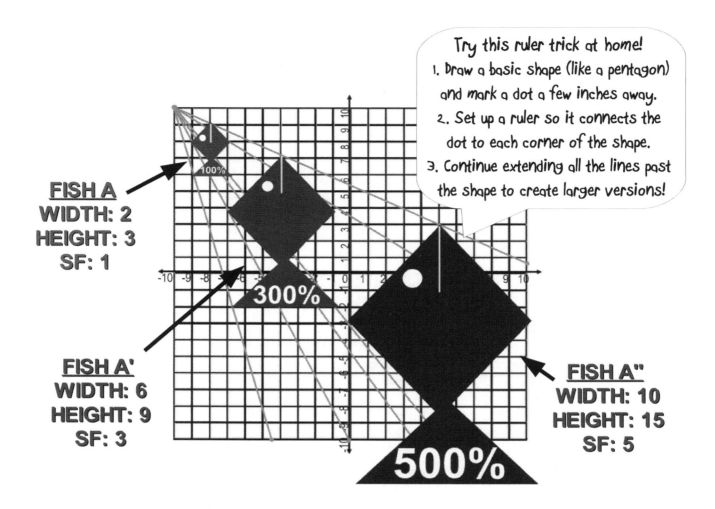

Having successfully reached this point, you are undoubtedly ready for a challenge. As a master of the **coordinate plane**, you can now learn some **graphing**, **symmetry** and **transformation** tricks that can be seen in **coordinate pairs** and on the grid as well. Take a minute to look at the fish **images** above. The point (-10,10) connects to each corner of **Fish A** in a straight line, creating larger **images**. What patterns do you see in the size of **dimensions** in each fish? What does the **scale factor** (SF) tell you about the size of the figures?

Scale Factor is the number that you multiply a **dimension** (like length, width, height) by to create a new **image**. For example, **Fish A** is 2 squares wide and 3 squares high. **Fish A'** (we read A' as "A prime") has a **scale factor of 3**, which means that it has been **dilated** (or enlarged) **to 300%**. Fish A has a side length **ratio** of 2:3, so **Fish A'** will have a side length **ratio** of 3 times bigger, or **6:9**.

Fish A" has been **dilated** to **500%**, so you will increase each dimension to **5X** bigger. The original **ratio** is 2:3, so the new ratio will be **10:15**. The ruler trick above works with enlarging dimensions, but you can use **scale factor** to enlarge specific **coordinate pairs** mathematically.

64

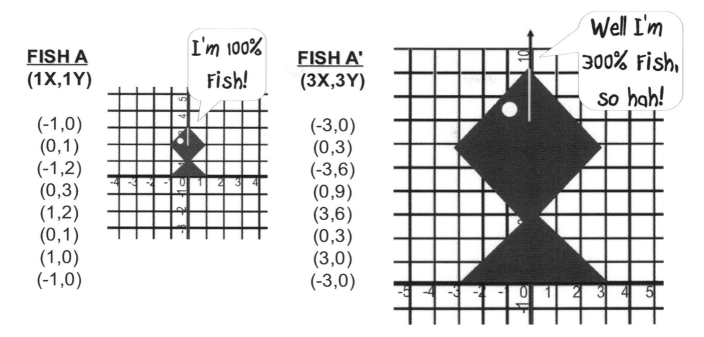

FISH A
(1X,1Y)

(-1,0)
(0,1)
(-1,2)
(0,3)
(1,2)
(0,1)
(1,0)
(-1,0)

FISH A'
(3X,3Y)

(-3,0)
(0,3)
(-3,6)
(0,9)
(3,6)
(0,3)
(3,0)
(-3,0)

Here's another look at **Fish A** and **Fish A'**. You'll notice that they have shifted, **or translated**, from high up in Quadrant 2, to right next to the **origin**. This will help us **enlarge** the fish using a **scale factor**. The tip of the mouth of **Fish A** *was* at **(-8,9)** on the previous page, but *now* it is at **(0,2)**. You can trace this movement with your finger, counting **right 8** lines and counting **down 7** lines. This can be stated mathematically as a **translation of (X+8, Y-7)** because the new image is shifted, and is the **same size and shape**. Identical **images** are called **congruent**.

Fish A' has also shifted from its previous position. His mouth *was* at **(-4,7)** but *now* it is at **(0,9)**. Trace with your finger, or use arithmetic, to write the **translation** to describe this shift. The mouth has moved **4 right** and **2 up**, so we write **(X+4,Y+2)**. **Fish A'** is *also* **300% bigger** than **Fish A**. Look at the coordinate pairs above for both fish. What do you notice? In each case, the coordinate pair has increased, or **dilated**, the same amount as the **scale factor**. (-1,0) multiplies -1x3 and 0x3, to become (-3,0).

Example 1: Dilate the coordinates for Fish A to create the coordinates for Fish A".
(500% bigger)

FISH A   **FISH A"**
(1X,1Y)   **(5X, 5Y)**

(-1,0)   (-5,0)
(0,1)   (__,__)
(-1,2)   (__,__)
(0,3)   (__,__)
(1,2)   (__,__)
(0,1)   (__,__)
(1,0)   (__,__)
(-1,0)   (__,__)

There is a reason why we cannot graph Fish A" on the paper provided in this book. What is it? _____

_____

Bonus Question: The image extends to +15, while the coordinate grid only has a range of -10 to +10. It would go off the paper.

Example 1 Answers: (-5,0); (0,5); (-5,10); (0,15); (5,10); (0,5); (5,0); (-5,0)

Before you dive in, let's practice translations, dilations & symmetry.

Example 2: Write the **translation rule** in **(X,Y) form** to get from one letter to the next.

Example: From D to A: **(X-7, Y+1)**

1. From A to B: (X+__ , Y+__)

2. From B to C: (X+__ , Y - __)

3. From C to A: (X - __ , Y - __)

4. From C to D: (____ . ____)

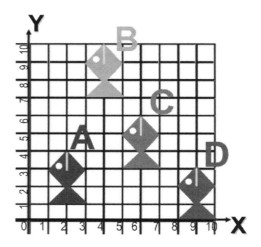

Example 3: Dilate and **graph** the images at right.

Triangle E: Rule (X,Y)	Triangle F: Rule (2X,2Y)	Triangle G: Rule (4X,4Y)
(1,0)	(2,0)	(4,0)
(1,2)	(__,__)	(__,__)
(2,0)	(__,__)	(__,__)
(1,0)	(__,__)	(__,__)
(Color yellow)	(Color green)	(Color gray)

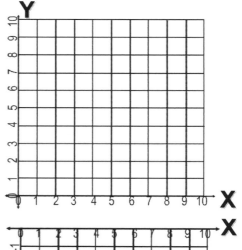

Example 4: Reflect and **graph** the images from Example 3 over the **x-axis** by changing each (X,Y) coordinate to (X,-Y). Will this create a **vertical** or **horizontal** line of symmetry? _____

Triangle E': Rule (X,-Y)	Triangle F': Rule (2X,-2Y)	Triangle G': Rule (4X,-4Y)
(1,0)	(2,0)	(4,0)
(1,-2)	(__,)	(__,)
(2,0)	(__,)	(__,)
(1,0)	(__,)	(__,)

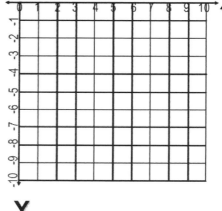

Example 3 Answers: F = (2,0); (2,4); (4,0); (4,-8); G = (4,0); (8,0); (4,0)

Example 2 Answers: F = (2,0); (2,4); (4,0); (4,8); G = (4,0); (8,0); (4,0)

Example 1 Answers: 1. (X+2,Y+6); 2. (X+2,Y-4); 3. (X,4,Y-2); 4. (X+3,Y-3)

21. STAR POWER

Connect each coordinate pair to the previous one, crossing off as you go.

In this puzzle, you will use the given rule to enlarge, or **dilate**, Bubba and Super Star in order to graph their **similar figures**.

Tiny Star

Tiny is the original size of all the stars.

Her rule is (1x,1y).

So all her points are 100% of their size.

Graph them like usual.

(0,3)
(1,1)
(3,1)
(1,0)
(2,-2)
(0,-1)
(-2,-2)
(-1,0)
(-3,1)
(-1,1)
(0,3)
Lift Pencil

(-0.5,0)
(0.5,0)
(0,-0.5)
(-0.5,0)
Lift Pencil

Add eye dots at:
(-0.5,0.5)
(0.5,0.5)

Bubba Star

Bubba has been dilated, or enlarged, to 200%.

His rule is (2x,2y).

So all his points are DOUBLE that of Tiny's.

Double the old ones, then graph!

(0,6)
(2,2)
(__,__)
(__,__)
(__,__)
(__,__)
(__,__)
(__,__)
(__,__)
(__,__)
(__,__)
(__,__)
Lift Pencil

Examples:

Now do the rest yourself!

(You can't graph Bubba's mouth and eyes because they are hiding behind Tiny)

Super Star

Super has been dilated, or enlarged, to 300%.

Her rule is (3x,3y).

So all her points are TRIPLE that of Tiny's.

Triple Tiny's points, then graph!

(0,9)
(3,3)
(__,__)
(__,__)
(__,__)
(__,__)
(__,__)
(__,__)
(__,__)
(__,__)
(__,__)
(__,__)
Lift Pencil

Examples:

Now do the rest yourself!

(You can't graph Super's mouth and eyes because they are hiding behind Bubba)

Make each star a different color when you are done!

Use 4-Quadrant Graph Paper

67

21. STAR POWER KEY

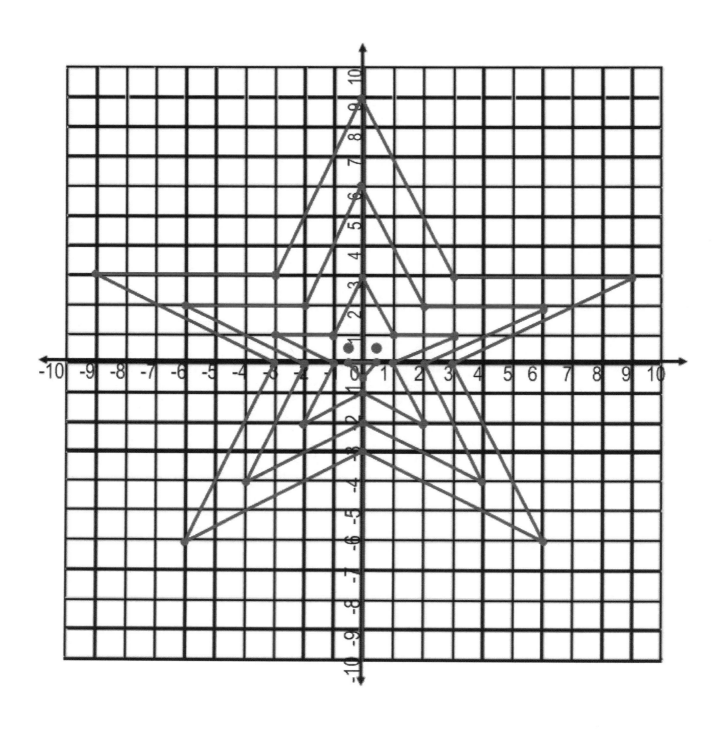

22. STAR SLIDE

Connect each coordinate pair to the previous one, crossing off as you go.

In this puzzle, you will use the given rule to slide, or **translate**, Tiny around Quadrant 1 to **transform** her into the images of her friends.

Tiny Star

Tiny is in the original position of all the stars.

Her rule is (x,y).

Another way to write her rule is (x+0,y+0).

Graph Tiny & color <u>yellow</u>.

(1,0)
(2,2)
(0,3)
(2,3)
(3,5)
(4,3)
(6,3)
(4,2)
(5,0)
(3,1)
(1,0)
Lift Pencil

(2.5,2)
(3.5,2)
(3,1.5)
(2.5,2)
Lift Pencil

Add eye dots at:
(2.5,2.5)
(3.5,2.5)

Slimy Star

Tiny <u>transforms</u> into Slimy by sliding 5 lines <u>up</u> on the y-axis.

Slimy's rule is (x,y+5).

So you <u>add 5</u> to all of Tiny's y-values, but x stays the same.

Graph Slimy & color <u>blue</u>.

(1,<u>5</u>) ◄— 0+5
(2,<u>7</u>) ◄— 2+5
(0,__)
(2,__)
(3,__)
(4,__)
(6,__)
(4,__)
(5,__)
(3,__)
(1,__)
Lift Pencil

(2.5,__)
(3.5,__)
(3,__)
(2.5,__)
Lift Pencil

Add eye dots at:
(2.5,__)
(3.5,__)

Grimy Star

Tiny <u>transforms</u> into Grimy by sliding <u>4 right</u> and <u>1 up</u>.

Grimy's rule is (x+4,y+1).

So you <u>add 4</u> to Tiny's x-values, and you <u>add 1</u> to Tiny's y-values.

Graph Grimy & color <u>green</u>.

1+4 ►— (<u>5,1</u>) ◄— 0+1
2+4 ►— (<u>6,3</u>) ◄— 2+1
(__,__)
(__,__)
(__,__)
(__,__)
(__,__)
(__,__)
(__,__)
(__,__)
(__,__)
Lift Pencil

(__,__)
(__,__)
(__,__)
(__,__)
Lift Pencil

Add eye dots at:
(__,__)
(__,__)

Examples:

Now do the rest yourself!

Use Quadrant 1 Graph Paper

22. STAR SLIDE KEY

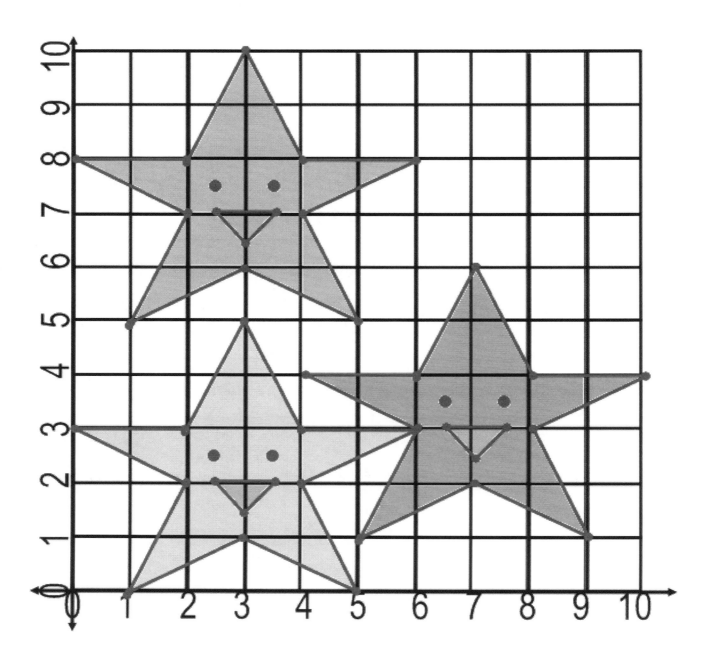

23. STAR SHUFFLE

Connect each coordinate pair to the previous one, crossing off as you go.

In this puzzle, you will use the given rule to slide,
or **translate**, Tiny around all 4 quadrants.

Tiny Star #1

Tiny is in the original
position in Quadrant 1.

Her rule is (x,y).

Graph Tiny as usual.

(1,0)
(2,2)
(0,3)
(2,3)
(3,5)
(4,3)
(6,3)
(4,2)
(5,0)
(3,1)
(1,0)
Lift Pencil

(2.5,2)
(3.5,2)
(3,1.5)
(2.5,2)
Lift Pencil

Add eye dots at:
(2.5,2.5)
(3.5,2.5)

Tiny Star #2

Tiny shuffles left 10
on the x-axis & up 4
on the y-axis.

Her rule is (x-10,y+4).

Graph Tiny in Quadrant 2.

(-9,4) 1-10,0+4
(-8,6) 2-10,2+4
(-10,7)

Examples:

Now do
the rest
yourself!

(__,__)
(__,__)
(__,__)
(__,__)
(__,__)
(__,__)
(__,__)
(__,__)
(__,__)
Lift Pencil

(__,__)
(__,__)
(__,__)
(__,__)
Lift Pencil

Add eye dots at:
(__,__)
(__,__)

Tiny Star #3

Tiny shuffles right 2
on the x-axis & down 8
on the y-axis.

Her rule is (x+2,y-8).

Graph Tiny in Quadrant 4.

(3,-8) 1+2,0-8
(4,-6) 2+2,2-8
(2,-5)

Examples:

Now do
the rest
yourself!

(__,__)
(__,__)
(__,__)
(__,__)
(__,__)
(__,__)
(__,__)
(__,__)
(__,__)
Lift Pencil

(__,__)
(__,__)
(__,__)
(__,__)
Lift Pencil

Add eye dots at:
(__,__)
(__,__)

Tiny Star #4

Tiny shuffles left 9
on the x-axis & down 7
on the y-axis.

Her rule is (x-9,y-7).

Graph Tiny in Quadrant 3.

(-8,-7) 1-9,0-7
(-7,-5) 2-9,2-7
(-9,-4)

(__,__)
(__,__)
(__,__)
(__,__)
(__,__)
(__,__)
(__,__)
(__,__)
(__,__)
Lift Pencil

(__,__)
(__,__)
(__,__)
(__,__)
Lift Pencil

Add eye dots at:
(__,__)
(__,__)

Use 4-Quadrant Graph Paper

Coordinate Graph Art for Grades 6-8

23. STAR SHUFFLE KEY

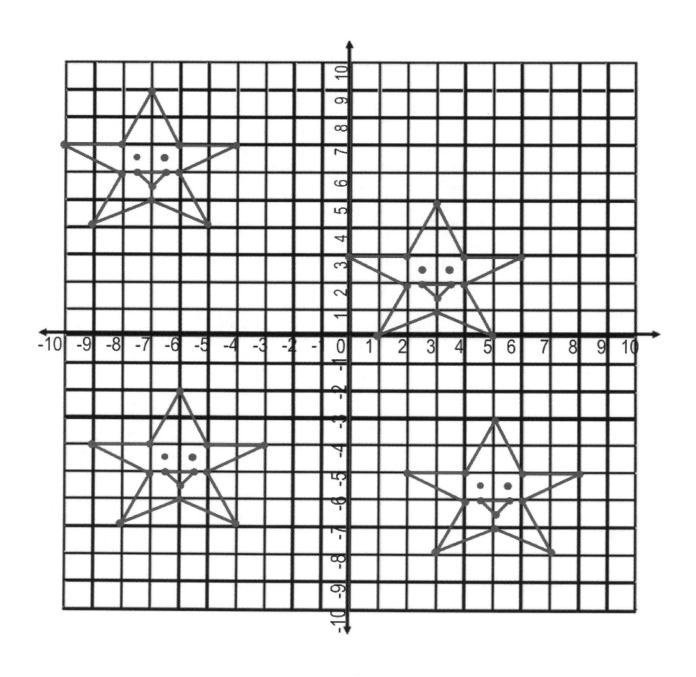

24. REFLECTING STARS

Connect each coordinate pair to the previous one, crossing off as you go.

In this puzzle, you will use the given rule to **reflect** Tiny Star like a mirror. First **horizontally** across the **y-axis**, and then **vertically** across the x-axis.

Tiny Star #1

Tiny is in the original position in Quadrant 1.

Her rule is (x,y).

Graph Tiny & color her <u>yellow</u>.

(1,0)
(2,2)
(0,3)
(2,3)
(3,5)
(4,3)
(6,3)
(4,2)
(5,0)
(3,1)
(1,0)
Lift Pencil

(2.5,2)
(3.5,2)
(3,1.5)
(2.5,2)
Lift Pencil

Add eye dots at:
(2.5,2.5)
(3.5,2.5)

Tiny Star #2

Reflect Tiny <u>horizontally</u> across the <u>y-axis</u> into Quadrant 2.

Her rule is (-x,y).

Fill in the coordinates & color <u>orange</u>.

(-1,0)
(-2,2)
(0,3)
(__,3)
(__,5)
(__,3)
(__,3)
(__,2)
(__,0)
(__,1)
(__,0)
Lift Pencil

(__,2)
(__,2)
(__,1.5)
(__,2)
Lift Pencil

Add eye dots at:
(__,2.5)
(__,2.5)

Examples:

Now do the rest yourself!

Tiny Star #3

Reflect Tiny <u>vertically</u> across the <u>x-axis</u> into Quadrant 4.

Her rule is (x,-y).

Fill in the coordinates & color <u>light yellow</u>.

(1,0)
(2,-2)
(0,-3)
(2,__)
(3,__)
(4,__)
(6,__)
(4,__)
(5,__)
(3,__)
(1,__)
Lift Pencil

(2.5,__)
(3.5,__)
(3,__)
(2.5,__)
Lift Pencil

Add eye dots at:
(2.5,__)
(3.5,__)

Examples:

Now do the rest yourself!

Tiny Star #4

Reflect Tiny <u>vertically</u> and <u>horizontally</u> into Quadrant 3.

Her rule is (-x,-y).

Fill in the coordinates & color <u>light orange</u>.

(-1,0)
(-2,-2)
(0,-3)
(__,__)
(__,__)
(__,__)
(__,__)
(__,__)
(__,__)
(__,__)
(__,__)
Lift Pencil

(__,__)
(__,__)
(__,__)
(__,__)
Lift Pencil

Add eye dots at:
(__,__)
(__,__)

Examples:

Now do the rest yourself!

Use 4-Quadrant Graph Paper

Coordinate Graph Art for Grades 6-8

24. REFLECTING STARS KEY

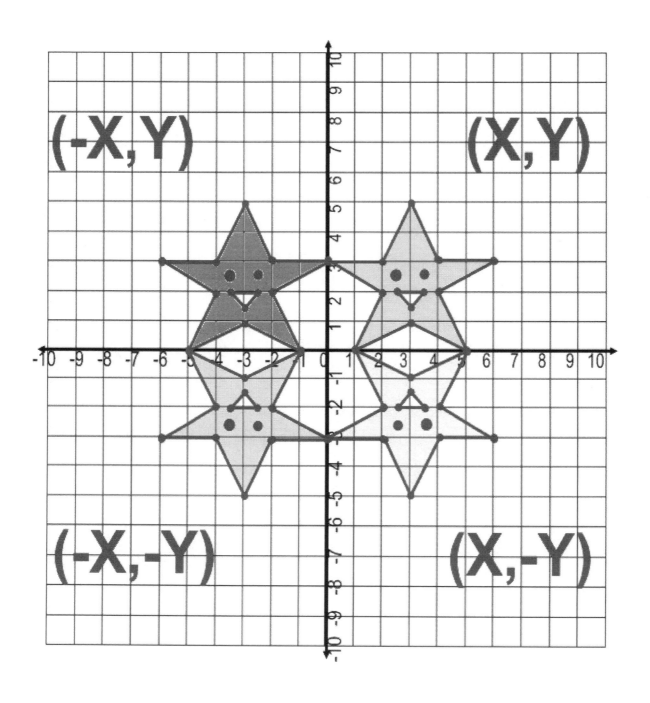

Coordinate Graph Art for Grades 6-8

25. STAR CHALLENGE!

Connect each coordinate pair to the previous one, crossing off as you go.

Bubba has taken over! In this **CHALLENGE** puzzle, you will try several _fraction_ rules to _shrink_ and _reflect_ Bubba back to Tiny's size.

Bubba Star	Spiny Star	Shorty Star	Tiny Star!
Bubba has taken over the original position!!	Bubba halves his x-values & reflects over the x-axis to create Spiny Star.	Bubba halves his y-values & reflects over the y-axis to create Shorty Star.	Bubba finally halves **BOTH** his x- and y-values. Oops, she's upside down!
His rule is now (x,y).	Spiny's rule is (½x,-y).	Shorty's rule is (-x,½y).	Tiny's rule is (-½x,-½y)
Graph Bubba as usual.	Graph Spiny in Quadrant 4.	Graph Shorty in Quadrant 2.	Graph Tiny in Quadrant 3.

Bubba Star

(0,0)
(2,4)
(-2,6)
(2,6)
(4,10)
(6,6)
(10,6)
(6,4)
(8,0)
(4,2)
(0,0)
Lift Pencil

(4,3)
(3,4)
(5,4)
(4,3)
Lift Pencil

Add eye dots at:
(3,5)
(5,5)

Spiny Star

(0,0) ½·0,-1·0
(1,-4) ½·2, -1·4
(-1,-6)

Examples:

Now do the rest yourself!

(__,__)
(__,__)
(__,__)
(__,__)
(__,__)
(__,__)
(__,__)
(__,__)
(__,__)
Lift Pencil

(__,__)
(__,__)
(__,__)
(__,__)
Lift Pencil

Add eye dots at:
(__,__)
(__,__)

Shorty Star

(0,0) -1·0,½·0
(-2,2) -1·2,½·4
(2,3)

Examples:

Now do the rest yourself!

(__,__)
(__,__)
(__,__)
(__,__)
(__,__)
(__,__)
(__,__)
(__,__)
(__,__)
Lift Pencil

(__,__)
(__,__)
(__,__)
(__,__)
Lift Pencil

Add eye dots at:
(__,__)
(__,__)

Tiny Star!

(0,0) -½·0,-½·0
(-1,-2) -½·2,-½·4
(1,-3)

Examples:

Now do the rest yourself!

(__,__)
(__,__)
(__,__)
(__,__)
(__,__)
(__,__)
(__,__)
(__,__)
(__,__)
Lift Pencil

(__,__)
(__,__)
(__,__)
(__,__)
Lift Pencil

Add eye dots at:
(__,__)
(__,__)

Use 4-Quadrant Graph Paper

25. STAR CHALLENGE KEY

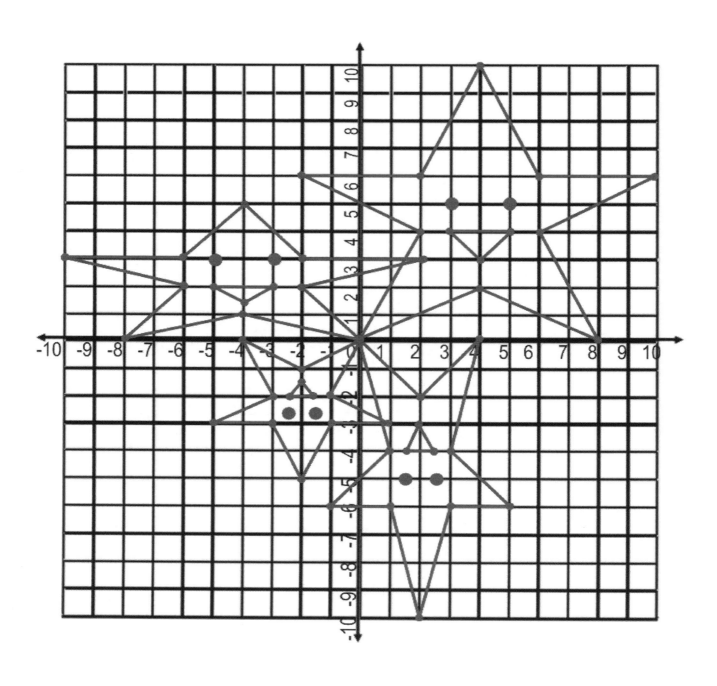

SECTION 6:
WEB RESOURCES

INTERNET-BASED INSTRUCTION

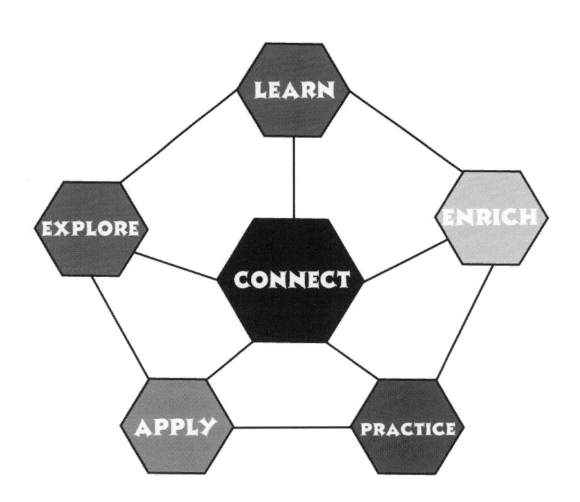

Name: _____   **Hour:** _____   **Date:** _____

WEB INVESTIGATION TOPIC:

_____

Website Name: _____   Minutes spent there: _____

What I did: _____

_____

Math Questions/Topics Comments: _____

_____

Website Name: _____   Minutes spent there: _____

What I did: _____

_____

Math Questions/Topics Comments: _____

_____

Website Name: _____   Minutes spent there: _____

What I did: _____

_____

Math Questions/Topics Comments: _____

_____

Website Name: _____   Minutes spent there: _____

What I did: _____

_____

Math Questions/Topics Comments: _____

_____

78

Web Resources: The Cartesian Plane

Website Name: "Introduction to the x,y-Plane (The Cartesian Plane)"

    Address: http://www.purplemath.com/modules/plane.htm

    Description: History of the coordinate system, building on one axis at a time. Good visuals, easy to read, thorough, with interactive applications.

Website Name: "Cartesian Plane" from About.com

    Address: http://math.about.com/od/geometry/ss/cartesian.htm

    Description: Four pages with printable images describing language and parts of the coordinate plane. The final page has practice items, and there are several links listed such as http://math.about.com/od/prealgeb2/ss/Coordinates.htm containing printable practice worksheets.

Website Name: "Graph Mole"

    Address: http://funbasedlearning.com/algebra/graphing/points2/

    Description: A silly practice game with three different levels of difficulty, requiring the player to select the correct coordinate pair from 4 choices, in order to hit the mole. Requires MacroMedia Flash Player.

Website Name: "Coordinate Plane or Cartesian Plane" Online Learning

    Address: http://www.onlinemathlearning.com/cartesian-plane.html

    Description: Contains four YouTube links to video tutorials, describing the functions of the Coordinate Plane, and giving examples. Includes many game and website links sorted by grade level.

Website Name: "Cartesian Battleship" on the full Cartesian Plane

    Address: http://www.green-planet-solar-energy.com/printable-math-puzzles.html

    Description: Printable directions and Battleship .pdf game boards for further instruction or remediation of material. Describes objectives and processes for learning.

Website Name: "Catch the Fly" interactive coordinate game

    Address: http://hotmath.com/hotmath_help/games/ctf/ctf_hotmath.swf

    Description: A wonderful game application for teachers with Smart boards or projectors. The bug flies around the grid, lands, and then a cute tree frog jumps up and shoots his tongue out and eats him!

Web Resources:
Transformations in the Coordinate Plane

Website Name: Kuta Software – Go to "Plane Figures" Section

     Address: http://www.kutasoftware.com/freeipa.html

     Description: Printable worksheets with attached keys, after my own heart! Contains practice for translations, rotations and reflections. Best to be used *after* the related lesson has been taught, as no instructional explanations are given.

Website Name: "Transformations in the Coordinate Plane" by Raquel A. Pesce

     Address: http://www.e-tutor.com/et3/lessons/view/54501/print

     Description: Mathematically precise explanations, diagrams, rules and vocabulary; great for students who like to learn on their own.

Website Name: "Coordinate Plane & Transformation Jeopardy"

     Address: http://www.superteachertools.com/jeopardy/usergames/Apr201015/game1271106514.php

     Description: Great sound and visuals for classroom review before a unit assessment or quiz. Includes options for teams, buzzer and "Final Jeopardy". Category topics are Glide Reflections, Rotational Symmetry, Reflection, Translation, and Dilations.

Website Name: "Translations, Reflections, and Rotations" from Shodor Interactive

     Address: http://www.shodor.org/interactivate/lessons/TranslationsReflectionsRotations/

     Description: One of the most complete sites on the internet for teachers, a lesson abstract and objectives are included, as well as standards, textbook connections, lessons and self-directed student practice. Click on the "Worksheet" or "Transmographer" tabs.

Website Name: "Alphabet Geometry: Transformations"

     Address: http://www.misterteacher.com/abc.html

     Description: Fantastic video clips that show transformations in progress, perfect for projectors or Smart boards. Click the "Advanced" and "Practice" sections for class tools.

Website Name: "Identify Transformations" by Online Math Learning

     Address: http://www.onlinemathlearning.com/identify-transformation.html

     Description: Contains four YouTube links to video tutorials, describing the different types of transformations, and giving examples. Includes game and website links sorted by grade level.

80

Web Resources:
Optical Illusions & Art on the Web

Website Name: "Transformations in Art"

    Address: http://www2.edc.org/mathproblems/problems/printProblems/ektransart.pdf

    Description: Printable worksheets that describe the history and use of transformations in order to create art, with practice problems and a key included.

Website Name: "Tessellations"

    Address: http://www.tessellations.org/

    Description: Extensive resources to the variety of transformations that occur in art around the world. Click on the "Tessellation" tab for information and examples. Also included is an "Art We Did" section and an "Art You Did" section where students and staff can submit their own artwork.

Website Name: "M.C. Escher – The Official Website"

    Address: http://www.mcescher.com/

    Description: Links to the famous artist's various symmetry, tesselation and translation prints. Click on "Picture Gallery" to look at his works in chronological order. Some images may not be appropriate for younger audiences.

Website Name: "Create a Translation Tessellation!"

    Address: http://www.jimmcneill.com/demo.html

    Description: Walks students step-by-step through directions and materials needed to make their own translation-based tessellation art.

Website Name: "Online Tessellation Creator"

    Address: http://illuminations.nctm.org/ActivityDetail.aspx?id=202

    Description: NCTM website resource for students to explore tesselations. Two drop-down menus allow students to read instructions and investigate which shapes tessellate.

Website Name: "Design a Tessellation Online – Escher Style"

    Address: http://gwydir.demon.co.uk/jo/tess/sqtile.htm

    Description: This is a great tool for beginners, with a grid already provided, and limited color and design choices. Students immediately see how their design will repeat, based on their shape choices. Click on "Escher-Style Examples" for inspiration. Students can also copy and paste the computer-based text language to save a picture.

81

Concluding Remarks

This book was created out of the desire to fill a gap. Graphing in Quadrant 1 felt to me like riding a tricycle, in that it was straight-forward and easy to succeed. Skipping right away to 4 quadrants, then, felt like handing a regular bicycle to a tricycle-proficient child, with no training wheels and minimal instruction, and saying, "Ride!". How many students can do that successfully on their second, third, fourth, or even tenth try? Probably not as many as we think. Thus, Sections 2 & 3 of this book were my attempt to "put training wheels on the bike". The benefits of improved coordinate plane awareness should be about as long-lasting as that of riding a bike. Many a branch in both Algebra and Geometry requires its use, as well as nearly all further study throughout high school and college.

In my ideal vision of this unit of study, each student would be pre-tested in order to place him or her in the correct section of difficulty. Many might choose to do the easier puzzles anyway, because they like to fill out check-lists (like the one provided), or because they love art. Others might only need 1 or 2 puzzles of choice before they feel ready to try the next section. Some may despise coloring the finished projects, while others may shade them in detail. Still others may race through the book and then have the freedom to explore on their own, with the web resources provided in Section 6 and the extension activities provided at the end of the Graphing Progress Tracker page. The end goal is to benefit and challenge every student, with minimal teacher instruction needed, unless she or he desires to do so.

In my short career of teaching in a variety of subjects, grade levels, and languages, and even shorter career in business, I have learned to value several things. The top of my list includes efficiency and continuous professional development. In my colleagues and students, I most value an acceptance and willingness to improve on weaknesses, while celebrating their strengths and building on them. Above all, I believe that a great idea is worth sharing. If this book has inspired you or your students, I would love to hear about it. If you have found your own instructional "gaps" that I might be able to assist with in a similar fashion to this book, I'd love to hear about that as well.

My professional website is: http://www.linkedin.com/in/mandybellm

Keep an eye out for future publications in similar student-friendly formats.

Anticipated topics may include:

Scientific Notation, Number systems & Measurement

2-dimensional & 3-dimensional measurement, including partial and combined figures

Smart Board applications & technology tips for the math classroom

New Teacher's Guide to the 21st Century Middle School

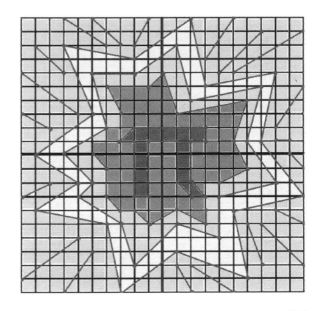

83

Coordinate Graph Art for Grades 6-8